The Kosmic Web

The Kosmic Web

Our spiritual and evolutionary
place in the multiverse

Peter Calvert & Keith Hill

attar‖books

First published in 2015 by Attar Books
Auckland, New Zealand

Second edition published 2022

Paperback ISBN 978-0-473-32411-7
Hardcover ISBN 978-1-99-115702-7
Ebook ISBN 978-0-473-32413-1

Cover photograph © Andre Stefanov, www.shutterstock.com

Attar Books is a New Zealand publisher which focuses on work that explores today's spiritual experiences, culture, concepts and practices.

www.attarbooks.com
www.keithhillauthor.com
www.wisdomschool.nz

Contents

Humanity is Bi-located

The Aura and the Electrospiritual

Agapéic Space and Shamanic Flight

Updating the Great Chain of Being

Epilogue 148

Preface

This is a channelled book. For many years we—Peter Calvert and Keith Hill— have been communicating with a group of non-embodied human beings who have completed their cycles of reincarnation on this planet. The Kosmic Web is (to date) one of twelve books in which these human identities offer their views on aspects of our existence. Each book has a particular focus. This offers a metaphysical overview of what, who, how and why we are. For us as collaborators, the book's primary virtue is the way it illuminates our over-arching metaphysical reality using current scientific and cultural concepts. This is reflected in the title, which uses the Greek word kosmos *(ordered world) to indicate that the book includes but reaches beyond the physical universe. We now stand aside so the guides may introduce this book themselves.*

WELCOME TO THIS LATEST PRODUCTION of our intent to share what we know with embodied humanity. To those who are first time readers of a book we have initiated, we send you a hearty greeting. To those who have read our books previously, we welcome back a dear friend. The purpose of this book is referred to several times in what follows so we will not elaborate on that purpose here. However, we will remark on the circumstances behind this book's production.

Three years ago we put into action a plan, which involves our two collaborators here, Peter Calvert and Keith Hill, to offer a refreshed approach to spirituality appropriate for those living in the twenty first century. The plan resulted in the production of *The Matapaua Conversations*. That book was the first iteration of a set of ideas that we have caused to be reorganised and expanded for this book. Most key ideas are contained in both books. But details vary. Each book also contains significant material not included in the other. So this book does not replace *The Matapaua Conversations*. Instead, each offers a different approach to the same overall set of ideas.

The plan to which we have referred was not concocted recently. It was set in motion long ago. Furthermore, it does not involve just these two but also many others, all of whom are contributing in their own ways, according to their particular strengths and capacities. In saying this, we wish to emphasise that individuals have not been picked out, and neither are those who are working with us special in any way. Each picked themselves, volunteering to contribute, in accordance with their abilities, to a much wider spiritual initiative to inform embodied humanity of the circumstances behind their current embodied situation. In this book we focus on two particular aspects of pertinent information. The first is the nature of the reality. The second is the source and purpose of human identity.

Elsewhere we have discussed our desire to create a new version of the *Bhagavad Gita*. That ancient work, like this, was initiated from the spiritual level to inform humanity of the essential spiritual aspects of their situation. That text, like this, deals with the nature of reality and human identity. That text, like this, offers a deep level of insight into the patterning of reality that facilitates the generation, growth, and knowing realisation of spiritual identities. That text, like this, was written in a form that was appropriate to the times in which it was produced. And that text, like this, as anyone who reads either will discover, requires intense study in order to extract the implications of what it contains.

This is a book to read slowly, to ponder on, to absorb. It is a book to come back to, to savour when you are in a contemplative mood. It is a book that we envisage having a slow pick-up among the wider human population. But those who absorb its intent will be deeply enriched by it, feeling their existence endorsed at a fundamental level. At the same time, we recognise that numerous contentious statements are made in these pages. Many of our statements can, and likely will, be misunderstood, misrepresented, even lambasted. That is the proper fate of books such as this, because if they did not offend at least some then they wouldn't be sufficiently robust in their confrontation of misguided, even ignorant, thinking.

Like *The Matapaua Conversations*, this book has been produced to correct misperceptions and misunderstandings. Accordingly, if something particularly strikes you, the reader, as beside the point or wrong or outright bizarre, we suggest this is a section you should focus on at length and ponder. Because it will most richly nourish your understanding. With that said, we begin by contextualising the source and purpose of this communication.

A Fresh Start

1

An Invitation to Become Excited

THE PURPOSE OF THIS BOOK is to add to those we have already initiated, the aim of which is to provide a clear description of the relationship between the spiritual and physical domains as they are understood by human beings.

These two domains are not separate. They are intimately connected. Of course, human beings do not perceive this, because humanity is naturally, and wholly appropriately, immersed in and engrossed by the stream of impressions provided by physical existence. Such immersion is appropriate because spiritual identities incarnate into human bodies so they may experience all that human existence offers—and human existence offers a great deal, even if individuals often feel the contrary.

However, among the many who live completely immersed in human experience, there are always a few who attempt to look beyond the physical everyday and seek to understand what is occurring on the other side of their experiences— that is, they seek a deeper, more spiritual understanding of what is happening to them and around them. This desire to understand human existence on more than a superficial level has led to the formulation of many different kinds of what may be termed meta-explanations.

The human need for meta-explanations

When you ascribe what happens in your life to fate, to chance, or to God, you are proposing a meta-explanation. Meta-explanations offer a non-physical rationale for what occurs in the physical world.

Historically, human efforts to generate deep level explanations have resulted in truths, half truths, twisted truths (a kernel of truth embroidered with fantasy), and complete untruths, along with every kind of permutation between. There are religious metaphysical explanations. There are scientific explanations limited to physical and sub-physical reality. And there are explanations generated by prophets, mystics, philosophers, artists and outsiders of all kinds, some straightforward, others complex, yet others quite bizarre.

Anyone who has delved into these deep matters will recognise that there is considerable variance of perspectives and great differences between meta-explanations, to the extent that many contradict others. As a result it becomes very difficult for anyone seeking a non-ordinary understanding of the world to decide which explanations are valid and which are not.

Some explanations appear valid, but only within a limited frame of reference. This can be said of religious and scientific meta-explanations, which exclude experiences that fall outside the boundaries of what they consider acceptable. One class of excluded experiences is the paranormal. The paranormal encompasses communicating with the deceased, having visions, perceiving others' thoughts, and undergoing out of body and near death experiences. Among the religious most such experiences are declared occult and associated with devilish intent. For the sciences, chemicals in the bio-system or errors in data gathering are used to explain away paranormal phenomena.

Religious and scientific meta-explanations are examples of when people narrow the frames of reference they use to make sense of the world. Alternatively, a frame of reference may be much more extensive than anything you have experienced. By way of examples, we offer accounts of religious prophets flying in chariots, shamans encountering nature spirits, and scientists tracing quarks. As a result of their perceptions, the prophet, shaman and scientist generate meta-explanations about very specific aspects of reality. But if their frames of reference exist beyond your experience, you are faced with the question of how you may best evaluate what is claimed. Do you simply believe what is asserted? Or do you reject it until you have direct confirming experiences yourself? This is problematic in the case of long dead prophets, South American shamans living in the Amazon, or the scientist ensconced in a high-tech lab. Simply, you cannot physically go to where they are to repeat their experiences. And anyway, you would have to engage in years of specialised study first.

Yet another difficulty compounds this issue. While a non-ordinary perception may be quite valid, the meta-explanation the experiencer offers may be much less so. To share their experiences such explorers have to convert non-ordinary experience into ordinary language. But in doing so, they may misrepresent what they experienced, either when describing it or in constructing a meta-explanation to contextualise it. This happens regularly in religious, scientific and so-called New Age contexts. The experience is valid. The meta-explanation is not.

These kinds of misunderstandings apply to all human-generated perceptions and their related meta-explanations, whether they are religious, scientific or imaginative in nature. The underlying problem is that they are all made from the perspective of an embodied human consciousness. But embodiment puts blinkers on human perceptions and cognition.

The bodily senses via which you perceive, and the brain's cognitive power that you use to make sense of your perceptions, are intrinsically limited. Your bodily senses perceive very little of what is actually occurring around you. And cognitively you understand very little of what has happened to get you here. This includes your own prior choices and how the environment you now occupy is structured to sustain the here and now of your life. To make this very clear, we repeat: all human meta-explanations are limited because the human vehicle you currently occupy is inherently limited, both in terms of your perceptual power and in terms of your cognitive ability to make sense of your perceptions.

So when human beings offer meta-explanations, some observations and explanations are inevitably valid. But they are intermixed with those that are partially valid, and others that are wholly invalid. And even those that are valid are valid only within a specific context and for a certain time. This last case is seen in the way that many historical religious, philosophic, scientific and mystical texts, which at their time of composition were viewed as revolutionary, are today perceived as interesting but outdated. Why does this happen? Because knowledge advances, cultures evolve, contexts change, and previous truths are supplanted by new truths. This brings us, somewhat circuitously, back to where we began.

The purpose of this book

To expand on our opening statement, our purpose here is to offer meta-explanations regarding how the spiritual domain is intermixed with and structures the socially conditioned physical domain occupied by humanity. However, first we need to say how our meta-explanations differ from human meta-explanations.

Where human beings need to break free of the constant flow of their everyday impressions, feelings and thoughts in order to perceive what lies beyond the everyday, we, as non-embodied identities, are already free. Where the human mind gains only brief glimpses of the non-physical domain, that domain is our normal habitat. And where the human mind is caught up in the minutiae of daily existence and has to struggle mightily in order to achieve any kind of overview of what exists, our perspective automatically offers us such an overview, both of human existence and far beyond what any human can perceive.

In asserting this, we make no claim for what human beings call God-like knowledge. This is not the case. We are just another form of spiritual identity, the same type as you, but at a different phase of our growth cycle. Our role is currently that of informant and guide. In human terms, you may see us as a much older and experienced sibling, or, if you prefer authoritative models, as a teacher. We remain nameless because it is what we offer, not who we are, that is significant in this exchange. But be assured that our nature is loving and our intention is to be helpful and illuminating.

To further reinforce that we make no claims to God-like knowledge, we acknowledge we have our limitations. Some exist by virtue of who and what we are. Some arise as a consequence of where we currently reside in that extended facet of reality that we call agapéic space, which humanity calls the spiritual domain. And some are limitations we have consciously adopted in order to fulfil our self-chosen role of passers-on of information on the inter-relatedness of the spiritual and physical. Yet, despite our limitations, our insights emanate from a level far beyond what any human body-encased mind accesses. This is where our understanding of the nature of reality most fundamentally departs from any human perception.

We state this not to make you feel small or inadequate, and certainly not to trumpet ourselves as superior in any way. Rather, we state it to stimulate excitement in you, the reader of arcane texts, which this book decidedly is. What if you could access trans-human explanations of reality? What if you could obtain explanations that are not bound by all the limitations of data and analysis that taint human knowledge acquisition? What if you could access a fresh spiritual-level way of viewing all that is? We ask these questions specifically because excitement at learning such things is what motivated our collaborator, Keith Hill, when he asked many deep questions of us, the result of which was the book *The Matapaua Conversations*. Similarly, a fascination with non-ordinary modes of perception, combined with an underlying attitude of daring-do, has motivated our other primary collaborator here, Peter Calvert, who channelled that material. While motivation on the human level inevitably waxes and wanes, these two have sustained their intensity sufficiently to get us to this point, at which we are collectively in a position to draw together a number of key concepts in order to present in this book a fresh overview, or what could equally be called new models, to describe the interlinked spiritual and physical domains.

However, before beginning this task we need to provide a little more detail regarding our perspective. We particularly need to make clear the wider context from which we are offering the meta-explanations that follow. In continuing these introductory remarks, we accordingly invite you to join us in an exploration of the nature of reality that is frequently not merely arcane but, from the perspective of the ordinary human mind, astonishing. We hope that the promise of such a revelation stimulates a requisite degree of excitement in you, our enquiring reader. With this clarified, we continue.

2

You are Part of a Greater Whole

THE PERSPECTIVE FROM WHICH we operate assumes reincarnation as a fact. We do not propose this after having analysed disparate sets of data and concluding that reincarnation is the most likely explanation. Nor do we propose reincarnation as a result of faith or metaphysical speculation. Instead, we do so on the basis of direct experience. We know spiritual identities repeatedly reincarnate in a sequence of human bodies because that is what we have done.

Our observation is that, on average, individual identities reincarnate one thousand times. Some reincarnate fewer times, others many times more. But one thousand is the average. As we have discussed the purpose and process of reincarnation at length elsewhere, here we will only sketch minimal details.*

A rationale for reincarnation

Repeated incarnations enable identities to experience, learn and grow. This means that no human being is what might be called a one-off, but is actually the singular sub-personality of an ongoing spiritual identity. Through adopting, entering and experiencing life via a sequence of sub-personalities, each ongoing spiritual identity progressively develops skills, derives knowledge, and accrues the ability to love in any and all circumstances. In this way, life by life, identities learn and evolve.

At the same time, each individual reincarnating identity is part of a larger entity, which could be termed a group soul. We are such a group soul. Our separate identities have completed their rounds of incarnations, having learned a phenomenal

---

* See *Experimental Spirituality*, *Practical Spirituality* and *Psychological Spirituality*.

amount from their many and diverse experiences. We have all since come together to form a newly re-assembled whole.

For the purposes of providing introductory background material, we add that we now exist at a particular frequency in agapéic space. (We will explain this term in due course.) From that position we offer insights to identities who are still incarnating. Given this is our relationship to the human domain, we would make two further comments.

The first is that this is why, in the previous chapter, we asserted that we are the same as you, that we are just at different phases of the growth cycle. We have completed the reincarnational process in which you are still engaged. All our insights into the human situation derive from that being the case. The second is that we are far from the only group soul currently offering humanity observations, advice, guidance, teaching and insights.

A range of insights is available

Other channelled material, emanating from other group souls, is widely available. It details the relationship of the spiritual to the physical in a variety of ways. This is as it should be, because different people learn differently and function best with different kinds of input. We offer particular kinds of knowledge and stimulate specific varieties of learning opportunities. Other reintegrated identities offer other kinds of knowledge and opportunities.

This variety occurs because during their incarnations some identities specialised in certain aspects of human experience, while others focused on alternative knowledge and information sets. This again is as it should be. As the saying goes, variety is the spice of life. By way of a metaphorical comparison, it could be said that just as there is a wide variety of plant species existing on this planet, each with their own qualities, so there is an even vaster range of spiritual identities existing in agapéic space, each likewise possessing different characteristics, interests, intents and means for achieving what they choose to do.

The result of this variety is that the types of information, insights and wisdom radiating from reintegrated identities may, and often does, differ significantly. It also explains why information from one source may be closely aligned with that from another source. This again is as it should be. Just as you have friends who are close to you, and other acquaintances who are not, so some at the reintegrated level are more closely aligned than are others. This does not imply that any kind of enmity or competition exists at the reintegrated level. Such behaviours are an entirely human, not a spiritual level, trait.

This is by way of introducing the next section of this introductory material. The circumstances that led to its original presentation are that Peter Calvert's meditation group is regularly visited by various kinds of non-embodied beings. Some we deliberately introduce to the group, to facilitate participants' learning. Other non-embodied beings see the lights are on (we mean this metaphorically, of course) and decide to drop in. Some arrive out of curiosity, others come with a purpose, yet others are not sure where they are or why they are there. Confusion is not the sole prerogative of the embodied human!

The following is a message from an identity who is aligned to our purpose and who arrived during the meditation group's regular sessions. Its intent was to pass on an affirmative message. That message is included here because it adds to what we have just stated regarding identity growth and reintegration.

A message on oneness

A significant stage of enlightenment occurs when an embodied identity becomes aware it is an autonomous individual, distinct from both its physical form and its spiritual form, that it is an individual component of a higher self, which is constituted of multiple components similarly derived from prior incarnations.

This perspective—of seeing oneself as a part of a larger whole—has historically contributed to human lore that describes human existence as an impenetrable mystery. In the past we have provided models so this situation may no longer be viewed as a mystery, but instead as part of a developmental process by which human beings acquire understanding. That understanding then contributes to the growth of the much larger whole that is an individual's higher self.

Individuals begin to develop a coherent understanding of their situation when they discern a sequence of events within a series of in-carnations. Peace can then prevail at the level of the ordinary embodied personality, given it fully appreciates that it has been here many times before, and in all probability will be here many times more. Using this understanding, the individual may conceptually extend the timeline of its existence beyond its present incarnation, which it now views as part of a series. This series possesses both a beginning and an ending. The ending necessitates a final sequence that involves a process of reintegration based on love. Reintegration occurs within the last rounds of embodied incarnation. It entails decreasing the difference between one part and another until there is no reason not to engage with every aspect of oneself. Every part becomes every other part. This is how unity is achieved. This also defines the condition of oneness.

Achieving the condition of oneness signals the start of a new phase, in which further knowledge and understanding are acquired at the level of a well-informed, capacious intellect—which would be deemed incredible by any living human being. Of course, other varieties of beings would view this extensive capacity as merely the restoration of a fragmented individual to its normal state.

In drawing attention to this, we emphasise that individual spiritual identities who have chosen to accommodate themselves within animal human bodies on this planet, and who regard their animal existence as normal, are actually engaging just a tiny aspect of themselves. The most intelligent individual who has ever existed on this planet is just an infinitesimal part of what it will eventually become. Intellectual pursuits, skilled capabilities, moral understanding and reasoning, and an accomplished artistic ability that makes the imagined and subtly perceived visible to others, are minor compared to the much greater capacities every single individual will eventually develop.

Thus we point to the eventual condition of each person present here, representative as you are of the world population as a whole. You, along with everyone else, are confined to a capacity far less than that possessed by the reunited, recombined, reintegrated identity that has achieved oneness.

Oneness is a long term goal. It may be aspired to, or ignored. It makes no difference which approach is adopted, because oneness is the eventual outcome for every person present here. For some, oneness will be with each other, given certain members present are aspects of one node of Dao-consciousness. Others are from different nodes of Dao-consciousness, within which they will achieve oneness. But the process is the same, irrespective of present identity.

We offer these hints of future possibilities in order to sustain a hopeful and loving atmosphere. We well understand that the present conditions are ephemeral, yet they are no less valuable for their short duration. We give thanks for the opportunity to be present with you, and now depart.

The nature and language of this communication

All the material on offer in this book has been communicated to you, the reader, via our two collaborators' auric channels. The auric channel is part of a wider structure of communication whereby information originating in the spiritual domain may be conveyed into the physically embodied domain. Our intention is to offer explanations regarding that wider structure of communication and how the human participates in it. We will presently discuss the auric channel by which this was communicated.

However, having such an intent brings us to the issue of what language is appropriate to this time and place. Our collaborators speak English and are the products of a high level of predominantly secular education. We say predominantly because both have studied a range of religiously-based spiritual traditions, historically, theoretically and, each to a certain extent, practically. They are both entirely comfortable with the conceptualising and experimental approaches of the sciences.

This is significant in the context of this communication, because the religious meta-explanations that once held sway in the West have fallen away and have largely been replaced by the naturalistic meta-explanations constructed by the sciences. This means that any communication such as this, which seeks to be accessible and relevant to a modern Western English-speaking spiritual enquirer, is advised to use the concepts and vocabulary of the contemporary sciences rather than those of pre-modern religion. That is what we are attempting here.

However—and as with everything to do with the human there is a further difficulty—just as in pre-modern times few possessed an acute understanding of religious and theological concepts and vocabulary, and so were unable to engage at a deep level with the arcane materials that underlay their own beliefs, so few in the contemporary secular West have intimate knowledge of all fields of the sciences or are familiar with all available specialised scientific terminology and theories.

In the book that preceded this, *The Matapaua Conversations*, the specialist notions initially came from the questions. The advantage in this was that the questions opened up opportunities for us to describe the processes that connect the spiritual and the physical using scientific terminology. The same applies in this book. Accordingly, in places the language will become somewhat specialised. So while we will attempt to keep our language accessible, illuminating arcane matters sometimes requires a specialised vocabulary. We apologise in advance, but it is the best we can do with the materials at hand. We trust you find the demands placed on you, the reader, will be worth the insights on offer.

This contexualising completed, we would make one last set of observations before presenting our new perspective of reality.

3

The Rationale for a Fresh Start

WE ARE TAKING TIME to establish our credentials, to use that business phrase, because we wish to indicate that we are not as strange as might be suspected, and that our motives are actually quite straight forward. Our intention is to share information that we have been asked for by various individuals on multiple occasions, and to do so in a way that is appropriate to the times and culture of those we are addressing.

As we made clear in the previous chapter, we are not alone in engaging in this business, as we might colloquially call it. Others are doing the same, with the same intent that we possess. Accordingly, we stand aside again and allow another identity to communicate through Peter Calvert. This identity is noted for its different location in agapéic space and for its particular energy signature, which when in immediate proximity to the human may be felt as serenity. Other adjectives that describe its energy signature include urbanity, wisdom and love.

On the need to decontaminate premises

We first need to identify ourselves, because introductions are the norm in polite interchanges. We are an integrated identity. As such we represent a condition of oneness. Oneness is a goal aspired to by many, who yet remain ignorant of its nature, what qualities it possesses, and the means by which they may acquire its condition. Accordingly, we will briefly describe our condition of oneness.

Our period of utilising human incarnation for educative purposes is complete. Incarnation for us has included many opportunities, many lives, many deaths, many interactions, and immersion in every human circumstance. We consider it is of no benefit to identify ourselves other than as having experienced all human life offers. For us, reintegration has since advanced to completion. The task of achieving unity has advanced to completion. Processing all accumulated information has advanced to completion. Accordingly, we can claim to have an integrated understanding of the role and purpose of being human.

Peter Calvert's availability as a communicative channel is convenient and, because he is disrespectful of convention, adequate to our purpose. We have consequently utilised his capacities in our best interests. Whoever finds our communication coincides with their own best interests are welcome to offer feedback in any way they choose.

That said, there are a number of things we wish to add. The first is to convey greetings to all humanity, and especially to those who elect to address what we intend. The second is to offer reassurances with respect to our intention. The third is to extend encouragement to individuals who have found that their life includes many challenges. Challenges are normal. They are the means by which an individual assimilates the human experience. We mean this in the narrow sense that it is through embodiment, and having to deal with all that embodiment entails, that an individual fragment comes to understand its role and purpose.

Issues around embodiment provide a specific theme of this transmission, which we understand has been progressing more or less smoothly, if intermittently, for more than two decades. It will continue through three decades, but probably not four. That is entirely appropriate, we could say even optimal, given many conversations are much shorter than that. Peter Calvert's colleague Keith Hill has also conversed at length with us. So we may introduce ourselves as being engaged in an exercise of parallel communication [involving both Peter and Keith].

We now need to comment on the degree to which a receiving mind plays an identifiable role in altering the message. Readers of channelled communications will naturally be concerned regarding whether the receiving minds in some way alter, modify, minimise, exaggerate, or otherwise materially add to messages on the gross level—or, more importantly, on the subtle and contextual level. We regret to affirm that the probability of that occurring is not zero. Nevertheless, we have put protections in place to minimise the opportunity for misinterpretation, biased phrasing, or for limiting this communication's perspective or scope. As a result, the material presented here has been engineered, one could say, to be of optimal quality.

We further observe that our intention for transmitting this type of communication is not to berate or cajole. Rather, we celebrate the knowledge that over the years has been brought to the lower minds of Peter Calvert and his friends and colleagues, united as they are in an endeavour to refresh spiritual understanding in the twenty-first century of your year-counting system.

We note there are many year-counting systems. None are better than any other. Year-counting systems are historically and culturally bound, having been manufactured for the simple purpose of coordinating an appreciation of the passing of time in the time-bound realm. That is their only purpose. We do not dispute the nature of time in any way. Time is a function of the domain in which you exist. It is not a function of the domain in which we exist. Yet we can access the time of our choosing. Which is what we are doing at this moment.

Finally, we support what has been transmitted over the years through Peter Calvert's higher mind to his lower mind, and affirm that he is not acting alone at either level.

An accelerated path to understanding

The reason individuals enter the physical domain is to embark on an accelerated path to understanding. That simple phrase, "an accelerated path to understanding", needs to be unpacked to reveal the depths of insight it contains. The fresh start this text offers is designed to offer precisely the required unpacking.

Of course, several centuries will pass before this is recognised. Nonetheless, we are taking this opportunity, at this moment in human time, to communicate via the particular combinations of higher and lower minds possessed by Peter Calvert and Keith Hill. We celebrate the plan these two agreed to prior to their embodiment, which is to project a dual voice into humanity, not a single voice, so any person who contemplates this material may appreciate a broader perspective regarding the nature and purpose of human existence. Their jointly operating processes ensures this understanding is distributed with a slightly higher reliability than might otherwise be the case.

The reason we emphasise this is precisely because, as we have previously discussed, we seek to decontaminate false premises, false products, and false interpretations regarding the nature of being human.

It is necessary to do so due to the capability—and willingness—of all social species to misconstrue their own nature, to believe in secondary products, and to accept their implications. Thus do social species deviate from their original intention and accumulate confusion and misunderstanding.

This is what happened to the intention conveyed through the founders of both Christianity and Buddhism. This is also what happens to the intention conveyed through all individuals who perceive that misunderstanding has accumulated in their community, and accordingly are driven by an urge to refresh their community's beliefs. Hence, this is another attempt, from our level, to establish a fresh, simplified, demystified, demythologised description of the nature and purpose of human existence.

Attempting to do so has consequences. At times the consequences are mortal, in the sense that the person generating a coherent and complex description of the role and purpose of human existence, which is naturally at variance with established tradition, is identified as a heretic and rendered mute by whatever means is available. The role of the martyr is not understood as sometimes involving this class of communication projected into the individual's mind. When the individual speaks out they may alienate and divide their community. They may then be viewed as so potentially damaging to the status quo that their removal from society is required. Removal may involve banishment, or death. Either way, they can no longer provide a correction to their community's false understanding.

We expect no such radical and demeaning shift in the life pattern of this individual [Peter]. His means of communicating our message is soft, in the sense of being passive and non-threatening. His is a small voice as far as his wider community is concerned. Moreover, he speaks in parallel with other voices, who similarly express the need for a change in understanding the nature of human existence. Considering that groundswell of change is well established, we see no reason not to add our voice. We are content to do so through this individual, because we believe the situation offers no risk to him. That leaves us free to articulate what we choose. Accordingly, we have chosen to offer these few ideas at this time, and will continue to offer others in the future.

This is an intervention in the culture

We wish to draw attention to the multiple nature of this set of transmissions. While it is not uncommon for two or more embodied individuals to be involved in a sequence of such communications, in this case multiple identities at the unified level are also involved. That is, this initiative is part of a larger intention to project from the spiritual domain into the human domain and achieve what may be termed "an intervention in the culture".

Such interventions have occurred frequently throughout human history, in multiple cultures and on numerous occasions. Christianity and Buddhism were just named as originating in two interventions of this kind. Even a superficial historical review of religions and spiritual traditions will reveal many other such interventions. Why do they occur? Because they become necessary. Humanity has the capability, and is frequently willing, to reduce a high level initiative to confusion and misunderstanding. Accordingly, new interventions are periodically required.

Such interventions cannot occur in a vacuum. They require opportunity. Opportunity in turn depends on embodied individuals who are primed to participate. And they need a social environment that is supportive of their activities, or at least does not hinder them. This is where pre-planning comes into frame.

As has been observed, these two individuals, Peter Calvert and Keith Hill, made an agreement prior to incarnation to participate in these transmissions. But they are far from the only ones fulfilling that intention. This means that given that throughout history interventions of this kind have been initiated repeatedly and frequently, and given that they are concurrently occurring today in both English-speaking and non-English-speaking cultures, there is nothing unique in what is happening here. In fact, from our perspective it could be said that the series of dual communications being made through these two is nothing more or less than business as usual.

What should be the appropriate response from your side of this communicative process? It is not to focus on us as the source, nor on the participating messengers. Certainly take note of the contextualising statements. But the best use of your time is to direct your focus and energy onto the material itself. There is more than enough in what has already been offered, and definitely in what follows, to occupy you for many years. That is, if you fully process the implications of this transmission and relate it the context of your life and its circumstances.

Having stated this, our introductory chapters are complete. As a concluding remark, we will add that an additional intent behind their production has been to "tune you in to our wave-length," to use that once favoured hippy phrase, and so prepare you for what it is to come. That achieved, let us now get down to the business at hand.

We start at the proper place, at the beginning. However, in this case it is the beginning of everything you, as an embodied human, currently know to exist. That is, we start with the beginning of the universe.

Kosmic Beginnings

4

The Origin of the Universe

ORIGINS HAVE ALWAYS FASCINATED human beings. Many wonder at their personal origins, who their forebears were, how they came to be the person they are. It is a natural human concern. Of course, underneath it, usually unperceived, is a drive by the individual's own spiritual self to make itself known to the personality it has chosen to co-associate with this time round. This is because anyone's ultimate origin is not familial, tribal, social or genetic. It is spiritual. The search for kosmic origins, the beginning of all that is, is an expression of that same drive emanating from your higher spiritual self to know where you came from.

The origin of the universe is well known to human scientists in a purely physical sense. The emergence of a singularity and the big bang that followed is an adequate explanation for what happened to bring this universe into existence. What is not well known is the spiritual intent behind that coming to be.

What we will attempt to describe is something of that intent and how it has created what is—for creation there certainly was. However, creation did not occur in the way that is traditionally and religiously understood. We will come back to the issue of creation presently. For now we need to clarify the nature and purpose of *this* universe, it being the universe humanity lives in and perceives.

The universe is in a multiverse

First what is required is the adoption of a field view. That is, you must picture an extensive field within which not just this universe but multiple universes are expanding and contracting. Science has called this the multiverse model. We affirm that this model is valid, although not in the way cosmologists currently speculate. As we will show, while both the religious and the scientific descriptions of the origin of the universe are valid in part, neither is a wholly successful description. There is much that both perspectives leave out.

How many universes are there within the multiverse? The number is only of theoretical concern to human beings, given no physical observation can possibly be made of another universe from this universe to validate what we might state. Nonetheless, we affirm there are a handful of universes currently going through their cycles of expansion and contraction, between five and ten in number. We will not be more precise for a reason we will shortly come to.

Each universe has specific parameters within which it exists. That is, just as scientists have identified certain numerical values in relation to this universe, such as the strength of gravity and the values of the weak and strong forces, and they know that these values are basic to this universe being what it is, so the parameter values basic to other universes are different, and consequently give rise to a different universal nature and conditions for existence.

Accordingly, it can be seen that the universe humanity exists within is actually part of a multiple initiative. It is not a sole attempt to conjure something from nothing, as the philosophers say. It is part of a wider experiment to explore many varied possibilities of existence. And we say *experiment* advisedly, because this is certainly one way the multiverse may be viewed. In fact, the model of the multiverse as an experiment is the model we prefer for this kosmic meta-explanation.

The multiverse is an experiment

In order to convey something of the nature of the multiverse, we projected an image of the multiverse into the mind of Peter Calvert. This indicates all the universes are connected within the field to an intent named the *observer*.

The *observer* is that which initiated the multiverse experiment. The *observer* selected the experimental parameters for each universe, then set them in motion. Given each universe is at a different phase within its complete repeated cycles of expansions and contractions, some will cease to be before others. They will, or will not, be replaced by other universe's according to the *observer's* intent. So this initiative of the multiverse is an ongoing process, the scale of which is beyond the understanding of any physical mind.

This explains why we cannot assign a particular number to the universes

PETER'S VISION OF THE MULTIVERSE

In this sketch, the uniform background is the featureless unmanifest and the spheres are the manifest multiple universes asynchronously and repeatedly expanding and contracting at their individual rate. That rate was approximately 1 to 3 seconds. Each universe manifested a similar maximum diameter before collapsing again. The web represents the tethering process of sustained intention to the *observer* in this experiment in the field of existence.

The implication is that in the same way that one would locate significant numbers of firecrackers somewhat distant from a house for simple reasons of safety, so also in that field of observation the undetermined behaviour of the universes would be placed at some distance from the *observer* for safety in the experimental setup.

Space seemed unlimited above, below and all around, being everywhere uniform in character. The *observer* was not observed by me but was known to be there. The individual universes seemed somehow tethered to the edge of the viewing field, equidistant in a radial arc from the locus of perception by the *observer*. I sensed my view was partial, but have no way of knowing what lay beyond the limits of my perception, although there were no other features visible in that scene.

FIGURE 1

within the multiverse. The human time scale is simply too truncated to accommodate the vastness of what is occurring. Humanity lives in a *now* that simply does not exist for us. Or for the multiverse. Or the *observer*. We exist outside the spacetime continuum humanity lives within. For us to generate a sample, which is what we would need to do, by making a theoretical slice through spacetime and then use that as a reference moment to definitively state, "As of now x number of universes exist within the multiverse," is ultimately an arbitrary exercise. There is no now. There is a range of current existence. From which we can derive an approximation only. Which we have done.

Of course, underpinning this entire kosmic construct is an assumption that the *observer* possesses the knowledge necessary to preselect values that result in a stable universe, which consequently functions according to the preset parameters, and does so in a manner that is consistent with the intent that brought it into existence. So this is not a randomised initiative. An infinity of universes has not been brought into existence out of a vague hope that one universe out of millions might succeed. This is definitely not the case. It is an intelligently intended and constructed experiment.

In this sense, the universe may be said to be created. Deliberately so. But we are not talking about some all-powerful God controlling all existence in the way religions conceive. We have no intention to do that. To expand on this we will discuss first the purpose of the experiment, then the nature of creation.

The *observer's* experimental purpose

Many scientists attempt to deny that there is any purpose to the universe and what exists in it, that everything happens accidentally, entirely as a matter of chance. They adopt this view because the only coherent alternative view is the traditional religious view, which is that God created everything. This could be a straightforward notion, but theologians have added to the idea of God as creator by also claiming that God is controller.

This notion of God as controller is difficult to sustain, because when anyone with an unbiased mind examines the universe they perceive that everything is not controlled, that chance and chaos are built into the universe's very fabric. So it is valid that the scientifically inclined reject the notion of God as controller.

However, because the religious have tied the idea of God as creator to that of controller, the scientific happily jettison the notion of the creator God too.

This is an example of the confusion human beings generate when they over-speculate. Neither religious metaphysics nor scientific naturalism are accurate. As a corrective to partial thinking we present the notion that the intent of the *observer* certainly led to the multiverse coming into existence in what could be said, using somewhat limited human terminology, to be a creative act. But no ultimate control is subsequently exerted in the traditional religious sense.

The *observer* has never been involved in any physical universe. Nor will it ever be. After setting the underlying parameter values for each universe, the *observer* has stood back, to use another inadequate human metaphor, and observes what occurs. This is somewhat like the Deist notion that a personal God created the world then stood back and let creation function according to natural laws. Yet the analogy works only partially. The *observer* is nothing like a personal God or a person of any kind. Instead, the personal functions at the level of nodes of Dao-consciousness. [This is explained in the next chapter.]

To come back to the origin of this universe, and to use current scientific terminology, following the big bang the laws of physics spontaneously transformed energy into matter, matter developed into gas clouds, gravitational forces drew the gas into stars and planets, and on certain of these planets environmental niches formed, some of which were supportive of life, in which life eventually emerged.

It could be said that behind the creation of each universe is a multiplexed intent to generate environments in which physical life, and ever more varied and complex forms of physical life, comes into existence and evolves. But throughout this extensive process there is never any intent, at the *observer's* level, to control either the process or the outcome. From the *observer's* perspective there is also no attachment to what happens to any single life form.

So the multiverse is an extended experiment of inconceivable magnitude in which sets of opportunities have been constructed for the purpose of facilitating the emergence of ever more varied life forms. There is benign goodwill at the outcome of each universe. But no outcome is predetermined.

The *observer* is part of the Dao

Accordingly, it can be affirmed that there is a spiritual level initiative to generate all the many domains of the physical. The spiritual existed before the physical. The physical is a sub-formation, a subconstuct, of a profound intent. The *observer* is the active manifestation of that intent. But behind the *observer* is an even deeper existent. We call this the Dao. We define the Dao as the ultimate unmanifest. The *observer* is the active force, for want of a better term, within the ultimate unmanifest. It too is unmanifest, of course. Simply, the Dao is the ultimate source of all that is and the *observer* is its intent.

We are trying to explain matters that exist at the edge, in fact beyond the edge, of human comprehension. Words are needed to explain anything. We have chosen the words Dao and *observer* because they are relatively neutral terms, at least in the context of secular Western English language.

There is a natural tendency among human beings to be either overly-respectful of traditional terminology or overly disrespectful. Neither approach is helpful. Many religious terms were invented in an attempt to explain arcane matters. Over time they inevitably accrued layers of theological dust, and their original intent became tarnished, then completely obscured. Many years later others have come along and rejected those traditional terms completely. Yet underneath the dust, under all the speculative and self-aggrandising verbiage, there remains a kernel of truth. We are trying to extract that kernel, polish it, and represent it from a fresh perspective using new terminology.

So if the term Dao, which is taken from the Chinese, or the term *observer*, which has connotations of the laboratory experimenter, conjure for some readers traditional concepts of God, we cannot stop them thinking in that way. However, we are deliberately attempting to avoid referencing religious God language. Aspects of what we are saying here have certainly been described before, in other cultures and eras, in religious contexts. We are merely developing and updating those older concepts, constructing culturally relevant models and using terminology appropriate to these times.

Accordingly, whenever we use the term Dao, we ask you to think of the ultimate, inexplicable unmanifest. And when we use the term *observer* keep in mind that you, with every sentient being, are also an observer. The creative initiative the *observer* undertook that led to the coming into existence of this universe, which ultimately generated the experiences that are your life, also exists in you. You can undertake creative initiatives and generate circumstances beneficial to yourself and others. You can set up experiments. You can bring into existence what previously did not exist—all in a far more limited way than the *observer* does, of course. But the *observer's* intent is not alien from yours. It is the same. Your desire to live a fruitful life is completely aligned to the *observer's* fecund intent to generate what has ultimately given you that opportunity.

We repeat, all this is an attempt to subvert the traditional view of God and the spiritual as separate and wholly distinct from physical human existence. This is quite incorrect. As we keep repeating, and as we intend to show through the remainder of this text, the spiritual and the physical are utterly and irrevocably intertwined. This starts with all physical universes being an extension of the intent of the spiritual.

That leaves us with just one topic to cover in this chapter. What results from the multiverse experiment?

The experiment generates information

All experiments result in the generation of data. In any experiment, data consists of the information that is generated both during the experiment and at the experiment's end, when the final result is reviewed.

The multiverse also delivers information at both these levels. As each universe reaches the end of its intended purpose, the *observer* collects all the resulting information. However, during the course of each universe's cycles of expansion and contraction much information is also generated.

Effectively, each universe functions as a laboratory in which vast numbers of experiments of greater and lesser scale play out across space, from the sub-atomic to the intergalactic scales, and for various durations of time, from the micro-second to the aeon. Within these ranges countless experiments of diverse kinds are occurring. One is the experiment that involves organic life on Earth.

After this planet first formed, conditions conducive to the formation of biological life led to life emerging. An experiment involving biological life has played out ever since, utilising natural evolutionary and emergent processes. The human animal, homo sapiens sapiens, is just one of untold species that have come into existence, settled into an environmental niche, adapted to survive in changing environmental conditions, and either evolved into new species versions or died out completely. We observe that this same type of experiment involving biological life is repeated on countless planets throughout the universe. This planet is far from the only one.

Within the overall biological experiment currently taking place on Earth, each species may be viewed as a sub-experiment. So the human animal species is a sub-experiment within the larger biological experiment. In fact, at present the human animal is a much too successful biological form. It has bred to such an extent that it is in danger of excessively polluting its environment and of out-breeding its food sources.

What usually happens when one species upsets the species balance is that self-regulation parameters present within the biosphere kick in, and large numbers within the over-bred species die out. Of course, the human species is extremely resourceful, creative and adaptable, which has enabled it to situate itself at the top of the planet's food chain. Given a tipping point is approaching, it is of interest to those observing, and of huge consequence to those living a human existence, as to how the human animal experiment will develop next. Just as no outcome has been predetermined for this universe within the multiverse, so no outcome for the human species has been preordained.

It may further be said that every single individual within a species is another experiment. This means that you, the reader of these words, are a experiment. Before you were born you set certain specific psychological, social, intellectual and creative parameters in place, according to which you now live this life. The experiences you have, the lessons you draw, and the conclusions you make, can be thought of as information drawn from the ongoing experiment that is your life. At your life's end you will acknowledge and process what resulted, extract further information from that result, and decide whether the experiment was successful, unsuccessful, or partially successful. You will then use this information to set the parameters for your next experiment, which will be your next human life.

This is exactly what the *observer* does in relation to each and every universe within the multiverse. Information is generated and collected throughout all levels of the kosmos. At the completion of each universe, the *observer* collects all the information. Who, then, does the collecting during the course of each experiment? The answer is spiritual identities like you. Each identity undergoes experiences in whichever domain they are located, spiritual or physical, collects information in relation to their experience, collates the information, and passes it up the next level for review. For the human, this next level is that of the group soul. The group soul is a node of Dao-consciousness. We will now explain what we mean by this term.

5

Dao and Nodes
of Dao-Consciousness

THE DAO IS THE ULTIMATE UNMANIFEST. Everything that exists derives from the Dao. Within the Dao we have identified the *observer* as the initiator of the experiment that is the multiverse.

It needs to be understood that the *observer* is not a personal manifestation of any kind. It cannot be associated with God. What we have attempted to express through the term *observer* is best thought of as an intent within the Dao. That intent, as we observed, is filled with goodwill, intelligence and love, plus many other qualities that cannot be expressed in human language because there is no equivalent experience in the human domain.

For example, the Dao in its intent is magnanimous and expansive in a way that human existence, focused on the much narrower concerns of the physical self, is incapable of manifesting. Even when the human spiritual identity is in a non-embodied state, which is actually its natural state, it would be overwhelmed if it was exposed to the expansive magnanimity of the Dao—and the term "expansive magnanimity" is a meagre attempt to capture the fullness of what the Dao is. Again we can only fail to make the inexplicable explicable. But the gesture is made.

So when thinking of the *observer*, think of an abstract intent that cannot be fully described. It is an urge. A momentum. That is all. Certainly, do not conjure in your mind the image of a being of any kind.

We began this chapter by asserting that everything that exists comes from the Dao, from the ultimate unmanifest. This includes all spiritual identities, which we are here naming nodes of Dao-consciousness. This is because, along with intent, consciousness is intrinsic to the Dao and derives from it.

Nodes are cast from the Dao

The Dao's nature is to manifest nodes. Nodes are individual agglomerations of consciousness that are cast from the Dao—by "cast from the Dao" we refer to the spontaneous natural development of a node that consists of the same substance as the Dao. Of course, the term *substance* is false, for the nature of the Dao has no substance in a conventional sense. So a node of Dao-consciousness may be thought of as a spontaneously agglomerated product of Dao-nature.

The arising of a node of Dao-consciousness is a spontaneous event triggered by a local accretion in density. A suitable metaphor is that of the ocean undergoing turbulent conditions, and as a result spontaneously emitting droplets from wave tips. This is a perfectly adequate metaphor, given that chaotic conditions in the sea generate forces sufficient to fragment the water, just as forces within the Dao generate something like tidal pressures. As a result, a variety of droplets come into being.

Nodes of Dao-consciousness are these droplets. Like droplets, nodes consist of various magnitudes and complexity. We will discuss the significance of nodal magnitude and complexity presently. But first we need to discuss what happens to a node when it first becomes manifest.

The state of the inexperienced node

When first cast from the Dao the node has little self-awareness. It can be barely differentiated from the unmanifest from which it is freshly accreted. Given it is of the same nature, it feels associated with the unmanifest, to the extent that it sees little difference between itself and the tidal forces which led to its manifestation. It is still bobbing on the wave, so to speak, by which it was cast from the Dao. Development is required for it to begin to identify itself as distinct from the infinite unmanifest.

Gradually, the node becomes aware of its situation. It perceives others of like nature around it. It becomes motivated to find out more about where it is and who it is with. It learns from those who are like it that various dimensions of existence are available for exploration. It also learns that it is possible for it to eventually make its way back to the ultimate unmanifest from which it was cast.

There is an underlay of excitement in learning this. It discovers it can explore. And it becomes eager to do so.

As the node manifests its intent to explore, others respond. So it finds community. Within that community is information. But the information is not easily understood by the node. The node doesn't know enough, and it lacks the inner resources to comprehend what it instinctively knows is available and may be understood. It becomes aware of this because it can see that many of those it is with possess extensive knowledge, love and wisdom. It feels frustrated. So now, having developed a strong sense of its own limitations, and being motivated to extend itself beyond those limitations, it seeks opportunities to learn and grow.

Accordingly, it enquires regarding the specific opportunities available to it so it may become loving and wise like others in its new-found community. It discovers appropriate opportunities have been mapped and thoroughly understood by those who possess greater experience. The community offers advice and counselling. This increases the node's awareness of what options may be preferred for self-development, which the opportunities are now seen as offering. By this means it is educated regarding what opportunities are most appropriate, and given the information it needs to make its first forays into learning.

Individuals at this stage are dew-drops, to use traditional terminology, conscious atoms, to use alternative terminology, novice spirits, to use a term from Spiritualism. They may otherwise be identified as inexperienced nodes of Dao-consciousness seeking their first association with a physical species.

How a node matures

When a node of Dao-consciousness co-associates with a physical species it inevitably starts to learn certain things. It becomes aware of its place within the order of life. This includes developing a sense of its relationship to other levels within populations of nodes. And it acquires information, knowledge and understanding, as a result of which its opportunities for taking on responsibility are enhanced.

It is through such experience that a node matures. And as it matures it becomes suited to explore more complex opportunities that are commensurate with its increased maturity.

A familiar model from the human world indicates this maturing process. An individual working in a large organisation can progressively ascend through the organisation, from the bottom to the top. As it does so, it acquires information relevant to each level. As it rises, it takes on more and more responsibility in relation to others, especially those directly under its care.

This exact same process applies to maturing nodes of Dao-consciousness. Although there is a proviso. At the human social level, as equally at the spiritual level, there are ranges and spheres of responsibility. According to the capacities a node possesses as a consequence of being cast from the Dao, so there are corresponding levels of responsibility to which it may aspire and eventually achieve.

Hence the specific qualities of each node's Dao-nature is reflected in what it aspires to in any environment, what kinds of information it seeks out and by hard work acquires, and how it processes that information to develop a mature understanding.

One other issue needs to be discussed at this point in relation to nodes. That is the issue of fragmentation.

The fragmenting and re-uniting of nodes

Not all nodes remain whole and complete throughout their existence. Some nodes spontaneously fragment after their emergence from the Dao. They do so for the same purpose that drives all nodes: to facilitate exploration and enhance their learning process.

Different nodes fragment into different numbers of fragments. There is a basic correlation between the size of the node and the number of its fragments. Larger nodes fragment into more nodes, smaller nodes into fewer. This is true as a general statement, but there are exceptions. There is no need for us to describe the exceptions as they are not relevant to human experience.

Fragmentation that *is* relevant to human experience includes those nodes that co-associate with the human, as well as with the horse and the cetacean family, consisting of whales, dolphins and porpoises. These nodes do fragment. In the case of human beings and cetaceans, it is the individual fragments that occupy bodies, sequentially, one body at a time.

As regards numbers of fragments, nodes that co-associate with the human species fracture, on average, into one thousand individual fragments. This number is approximate. Some human-related nodes fracture into fewer fragments, some into more. Nodes that co-associate with horses and cetaceans fracture into fewer fragments than the human. We will deal with these nodes and their related species elsewhere, as it falls outside the purview of this text.

Do all fragments of a node select to incarnate in the one species? Generally, yes. This is in part because they all receive the same advice from those who possess the same quality of Dao-nature. It is also because they wish to engage in their cycle of incarnations with those they know, these principally being other fragments of the same node, or fragments of other nodes closely associated with their own.

Nonetheless, some fragments do choose to co-associate with more than one species. And some select to co-associate with species on other planets completely, whether in this or in other galaxies. They may even explore opportunities offered by a species existing in a different universe. Selection is made entirely on the grounds that what is chosen offers useful and appropriate experiences. Given each fragment has choice, all this is possible. And so, inevitably, some choose to do so. That includes some who also choose to occupy human bodies.

Each node of Dao-consciousness possesses all the qualities of the Dao. These qualities include identity, intellect and purpose. Consequently, each individual fragment of a node also possesses identity, intellect and purpose. So you who are reading this transmission are a fragment of a node of Dao-consciousness. You have individual identity, intellect and purpose. You have self-creativity and choice. You have utilised these qualities to shape all your prior incarnations for the purpose of experiencing and learning. And you will continue to utilise them to evolve according to your self-creative intent.

Fragmentation leads to another phenomenon that applies only to nodes that fragment. This is that at the end of their maturation cycles all of a node's fragments come back together to reunite and form what was earlier called the group soul. In the context of what we are now discussing, it can be said that the group soul is a reintegrated node of Dao-consciousness.

After all a node's fragments have matured sufficiently, they are designated as having completed their cycles of reincarnation. They then reunite with every other fragment from the same node. In doing so, each one brings back everything they have experienced and learned, every skill they have developed, the wisdom generated from every responsibility they have taken on and successfully negotiated. As a result, the node is incalculably enriched in comparison to its initial inexperienced state.

Underlying each exploration is the drive for enrichment. Enrichment of the individual. Enrichment of whatever environment a fragment occupies to which it can contribute. Enrichment of the node when the stage of reunification is reached. And enrichment at other levels beyond that of the reunified node, levels we will not discuss here. The process of enrichment goes all the way back to the Dao, to which everything is eventually returned.

The image of a node emerging from the Dao, spontaneously fragmenting, and its thousand or so fragments then independently spreading out through the kosmos to explore, experience, interact, learn and evolve, could sound to some like a process a writer of science fiction might concoct. Nonetheless, it is true.

The physical multiverse is your domain. The expansive dimensions of spiritual space are your domain. Embodied realms, non-embodied realms—they all present opportunities for inquisitive nodes and their questing fragments. You are currently bound to a physical form. And you have definitely made a commitment to co-associate for a time with the human species in order to work through the conditions and experiments you yourself have set in train. But this will not always be so. When you take your place within a reunified node of Dao-consciousness, the kosmos in all its variety and vastness will await your engagement. The experiment is that vast and that fascinating. And it is as exciting as you choose to make it.

6

Identifying and Seeding
Environmental Niches

AS EACH NODE MATURES it develops more resources within itself. It is therefore able to engage in correspondingly more demanding experiences and tasks. As we stated in the previous chapter, fragments of a node of Dao-consciousness need to be prepared before they take on the difficult task of occupying the body of any species. Preparation facilitates opportunity.

We also note that co-habitation within a body, involving full immersion within bodily experience, is not the only way for a node to co-associate with an animal species. Some nodes remain non-embodied while overseeing and being responsible for individuals within a species. Among these nodes are those that associate with animal, insect and plant species on Earth. Some nodes have responsibility for just a few individual bio-identities at a time, some for many. Because different nodes have different capabilities, they find ways to associate with species that match their capacities. We will discuss this in detail presently.

The main point we are making here is that the human situation is just one among a range of possibilities by which nodes associate with the physical domain and with biological creatures within it. Yet another way of associating is to adjust physical forms directly from the spiritual domain. This is what occurs during the course of finding and seeding environmental niches.

The seeding of environmental niches

The environmental niches that are present throughout the multiverse offer two independent sets of opportunities for nodes of Dao-consciosuness. They offer opportunities for a node to directly associate with individuals of a species, as we have just discussed. But before a node can co-associate with a species an appropriate species has to be found. So a second range of opportunities involves node finding and developing suitable physical species, with which other nodes may usefully co-associate.

During the search for suitable species a long-term view is usually adopted. By this we mean that environmental niches are identified and marked as a possibility long before a species suitable for co-association actually occupies that niche. Often nodes adjust the environment, or the primitive species living in it, at the level of their DNA. It is a process of creative adjustment within existing physical and biological parameters. That is, the nodes work with what is available. They do not alter the fundamental physical and biological bases of the life they find. They work within what is already there.

We add a side observation here that there are nodes who engage in a higher level of creativity, doing the galactic equivalent of what human beings term terraforming. They can and do adjust planetary environments at a very fundamental level. We will discuss this process and the nodes that carry it out another time, in another context apart from this.

To return to our discussion of environmental niches, when opportune locations are found to have formed, and when they are seen to provide well-suited niches for species to exploit, they are seeded. At times this extends to organisms being created and then being left to their own devices. Thus nodes of Dao-consciousness engage creatively with environments and the species occupying them throughout this universe and every other universe. This is done responsibly and with goodwill, for the sake of their fellow nodes. It additionally expands and enhances the capacities and lives of the bio-entities they influence.

There is nothing unethical or untoward in this. It is precisely what human beings do when they plan, plant and oversee a garden, including when they interbreed and genetically alter plants to achieve a certain colour or shape, or for nutritional or hardiness-related outcomes. Except that, at the spiritual level, there is no equivalent to the weeds that worry human gardeners. At the spiritual level, creativity involves nurturing, enhancing and planning to facilitate long-term objectives, that is all. But in achieving these objectives, and in harmony with the intent that is the *observer*, control is not involved. Species are neither coerced nor accidentally killed off, as an unforeseen by-product of an ill-researched decision, in the way human beings do such things. Sometimes a predator species may be encouraged to shift away from an environment in order to facilitate the population growth of a target species. A species may even be "weeded out", but this is rare. More usually there is an adjustment, an intervention, at a particular, generally early, stage of species' evolution, then it is left to follow its course.

The extent to which these experiments have been initiated and carried out throughout the multiverse is beyond the scope of any individual mind to encompass. So at this time we can only broadly indicate the scope of it and leave it at that. But we can affirm that the process is dispersed across this and other universes, that it is always ongoing, and that its primary purpose is to positively and fittingly influence emerging environmental niches, preparing them so they may eventually be occupied by nodes possessing intellect and purpose. All this occurs within and in accordance with the experimental parameters with which the universe was created in the first instance.

To conclude: Various nodes recognise and explore an opportunity to exercise their creativity in the formation of life forms appropriate to existing environmental niches. There is no certainty as to which niche can or cannot be exploited. And there is only a probability of survival after a particular organism has been matched to a specific niche. Therefore the opportunity is both for initially seeding of environmental niches with organisms, and for reseeding it when it has been experimentally determined what capacities an organism requires to suitably fit the niche and to satisfy nodes' co-association requirements.

Having discussed nodes and niches in general terms, we will now focus on what has happened on this planet.

Identifying suitable niches on Earth

Environmental niches exist in various places across the Earth, for short or long periods. Speaking in the context of the long historical past, after a suitable environmental niche was identified from the spiritual level nodes took responsibility for seeding it with appropriate organisms. Organisms were created, or modified from preexisting stock, and distributed in a way that breeding populations became established. They were then left to follow their own courses. Species that could survive, did. Those that could not went extinct. There was never any particular concern over either outcome. It was the relationship between the characteristics of the niche and the surviving organism that mattered and was recorded.

This type of activity is part of an experimental imperative, which manifests as a creative drive, that all nodes of Dao-consciousness possess. The reason this creative drive was deliberately embedded in emergent nodes of Dao-consciousness is clear—it was so the Dao itself could explore the consequences of physical life as it emerged, evolved, matured and came to the end of its cycle. And it enabled engendered nodes, which are of itself, to participate on all levels in the process. These levels are from outside, observing from a distance; from near, participating by adjusting physical processes; and from inside, directly experiencing all that occurs.

Originator, outside, near, inside, re-absorber—this is the mystery of existence in which everyone and everything, from the smallest microbe to the largest Dao-entity, are collectively engaged. The Dao is examining, experiencing, appreciating and extra-producing itself through all nodes, including us. Meanwhile, the whole kosmos is shifting, fracturing, kaleidoscoping, reforming and evolving.

This is the wonder of what is!

Human embodiment on Earth

The process of node fragments co-associating with the human species began not long after the development of the species homo sapiens sapiens. It is an occupancy that was prepared a long time before. Pre-human species, humanity's ancient forebears, were initially seeded by nodes from the spiritual level, then left to follow their natural biological course. When they were ready to be occupied, that was done.

Occupancy began when it was observed that a sufficiently large breeding population had become established that those overseeing the process were confident it would survive as a viable species. Naturally, it took several million years to reach that point, given the entire process began with the initial evolution of biped animals, followed by their development into a distinctive line of species. When, after observation, the decision was made that homo sapiens sapiens was suitable for embodiment, and that the process of embodiment would be able to continue into the foreseeable future—which at that stage comprised the order of ten million years—the process of embodiment began.

The precise prognosis was that the species would allow not only the development, through continued embodiment, of individual history and karma, but there would also be sufficient time for all the embodied to resolve karma. Accordingly, an individual node of Dao-consciousness could be confident that specialising in this particular species would be adequate for the completion of their purpose, which is to achieve maturity and no longer require embodiment in an organic species. On the basis of this projection, the decision was made that the species was viable.

Were the bodies of any other species used for embodiment before homo sapiens sapiens? Yes. Which species were used? Several. Denisova hominins, homo floresiensis, homo sapiens neanderthalensis, and others. But not homo erectus itself, nor homo habilis, as they existed before a sufficiently positive projection of the survival of the proto-human species was developed.

Were the great apes used for embodiment? Yes, to some extent. Those species were trialled to decide whether they provided the required opportunities to express intellect and purpose, and to provide a process of migration across bodies that had sufficient depth to carry and complete karmic cycles. However, it was concluded that it would be more appropriate to access a different time period, rather than when the great apes were the primary occupants of the available environmental niches. It was foreseen that metabolic and intelligence adjustments could be made to that species in such a way as to generate a sub-species that offered more suitable opportunities.

Bearing in mind that biological adjustments have been made at the level of DNA, it may be said that suitable species have been bred from the available stock. As we have said, there was an intervention in order to increase the probability of a viable proto-human species suitable for co-association with nodes of Dao-consciousness.

We conclude these observations by venturing even further into what many readers might already be considering ungrounded speculation. We maintain that the human animal's cranial capacity was deliberately enlarged during the species' evolution. Part of the reason it was done was to accommodate a more complex brain. This enabled a more sophisticated, alert and aware mind to capture and process a greater order of information and to analyse it at a suitable rate.

An inevitable consequence of incarnation is the quantities of information being traded, as it were. They are mind-bogglingly huge. Most individuals do not comprehend the depth of analysis that accompanies and drives every daily activity that is not a habituated action. However, this is the result for which incarnation is undertaken. This is the promise and the possibility of incarnation, which every node of Dao-consciousness dives into physicality to obtain. It is the gold, the nugget, the pearl beyond price. It is the Dao's goal, which animates each and every willing node.

Humanity is Bi-Located

7

Dao-Consciousness
+ Bio-Consciousness

ALL LIVING BEINGS POSSESS consciousness. Consciousness is generated when the Dao becomes focused and generates a node that possesses the same qualities as the Dao. Earlier we described these qualities as identity, intellect and purpose. To them we now add a facility to respond to external input and a self-creative attribute that enables nodes to use input nutritiously to develop. That is, as they accumulate experiential information, and as they process it, they grow. They become greater than they previously were. They evolve.

Given that all nodes eventually redistribute what they experience and process back to the Dao, it may be said that the Dao is evolving with the evolution of each of its countless nodes.

The other introductory point we would make here is that consciousness is ubiquitous. Consciousness is everywhere. Vast numbers of nodes are exploring and creating throughout incalculable regions of spiritual space—which we are calling agapéic space—and equally throughout all the regions of physical space that extend through the multiverse. The physical is actually a small portion of the entire kosmos. This is difficult for human minds to conceive, but there are more nodes in existence than there are stars in all the multiverses. Many, many, many more. The magnitude is inconceivable to the human mind.

In the context of the multiverse, it must be appreciated that the penetration of the Dao is universal. Therefore consciousness is present within every being existing, whether spiritual or physical. Accordingly, a variety of living beings, on the level of species, also possess identity, intellect and purpose in varying degrees. We will explain.

Two levels of consciousness

As we have proposed, spiritual identity exists as a node of Dao-consciousness. A natural function of consciousness is to process information. Each node use its intellect to process what it experiences and to extract from it all kinds of information. It subsequently adjusts its purpose according to what it learns.

As a node explores the kosmos it enters unfamiliar environments. These are previously unknown zones occupied by unencountered identities. Remember, we are not referring to physical environments and identities here, but to spiritual identities exploring agapéic space. During these explorations many sensations of difference and similarity are felt by the node. Processing how others who are encountered are different or similar—some will be both—generates information. Further information is generated when the node acknowledges its own internal response to what it encounters. Even more information is generated when it adjusts itself in response to the difference and similarity it has experienced. All this becomes information accrued by the node as it negotiates the kosmos. Obviously, this is a very simplified and abstract description. But it makes the point: each node's experiences generate layers on layers of spiritual level information.

What nodes of Dao-consciousness do in agapéic space is also done by all species in physical space. As members of biological species interact with their local environment, they also encounter difference and similarity. These translate into such things as food, which is different from the creature but of sufficient similarity to be edible, or into identifying other creatures as being of the same or different species or family. Deciding whether other creatures are threats or friends, and responding appropriately, is crucial to each creature's survival. So is learning from experience. All this requires the creature to process perceptual input, for which it uses its nervous system and brain. As was stated at the end of the previous chapter, throughout any biological creature's life it generates a mind-boggling volume of information. This includes what is derived from its interactions with its local environment, with other creatures in it, and from its internal responses to what happens during foraging, mating, nesting, procreating and dealing with threats and the claims of kinship.

Hence, if it is maintained that a function of consciousness is to process information, and given that the biological brain indisputably processes masses of information, then it is valid to conclude that the biological brain possesses consciousness. So just as we have identified Dao-consciousness, so we equally now identify bio-consciousness.

The function of consciousness at the level of a node of Dao-consciousness includes focused awareness, observational capacity, identity, intellect, purpose, the ability to initiate various kinds of action to achieve a purpose, and a self-creative capacity to direct experiences, process them, learn from them and consequently evolve. Consciousness at the level of the biological brain is the product of a physical information processing unit, an organic nerve centre, whether it is primitive, as in the case of an ant, or advanced, as with the human brain. Like a node of Dao-consciousness, the biological brain is aware, perceives, generates identity, functions with purpose, has intellect to process perceptions, initiates responses, and learns from its experiences.

Moreover, just as there are inexperienced and naive Dao-conscious nodes and experienced and mature nodes, so there are inexperienced and naive bio-conscious creatures and experienced and mature creatures. Accordingly, we propose that identity be divided into Dao-identity and bio-identity, intellect be divided into Dao-intellect and bio-intellect, and acknowledgement be made that there is Dao-level purpose and bio-level purpose.

Innumerable varieties of all kinds of consciousness extend throughout the kosmos. And all have equal validity. Because they are each an expression of Dao.

Differences between nodes

We observed in an earlier chapter that as nodes of Dao-consciousness are cast out of the Dao they are not all engendered the same. There is variety among them. We will now discuss this further.

Nodes of Dao-consciousness exist in a range of sizes and complexities. There is no simple way of explaining this, except to compare nodes to drops of water and say that some drops may consist of just a few molecules of H_2O, as is the case with fog, while others are quite large, like the drops that accumulate on the end of a stalactite. The drops are larger because they have more water molecules in them.

Translating this metaphor back to nodes, some nodes are larger because they have more Dao in them. We are still speaking metaphorically, of course, because the Dao is not a substance. But it could legitimately be said that larger nodes possess a greater magnitude of Dao-intellect, purpose and nature. A simple way of explaining this is to say that larger nodes are more complex than smaller nodes. We need to make clear there is no valuative agenda concerning this difference. Larger is not better than smaller. They are just different. In practice, every node will, at some time in its explorations of agapéic space, encounter nodes that are less complex than it is and others that are more complex. Some are incomprehensibly larger and more complex, and some are fascinatingly tiny. But just as you do not judge and reject a flower because it is small, especially given some tiny flowers are overpoweringly beautiful, so a node does not place valuative judgement on other nodes based on magnitude or complexity.

Of course, there may be some nodes a node prefers to remain in the vicinity of, and others it prefers not to. But that, as is said on this planet, is another story.

Nodes use physical species to evolve

Nodes of Dao-consciousness naturally seek to experience, learn and evolve. Entering physicality offers opportunities to do just that. Accordingly, nodes co-associate with physical species for their own advantage. Extending their awareness into the physical domain enables them to experience situations that are not available in the spiritual domain. For example, being in mortal fear of your life is not an experience that it is possible to have in the purely spiritual domain of the Dao. Nor is self-sacrifice, such as parents willingly do for their children, friends do for each other, or citizens do for their country. On the other hand, many other experiences do translate between the spiritual and physical domains, such as being creative, nurturing others, coming to understand others, and fostering environmental niches.

It could be said that just as the Dao's intent has created the multiverse as an extended experiment, setting up parameters then allowing them to play out in whichever way they do, so each node chooses a physical species to co-associate with, selects certain parameters that broadly define the nature of its possible experiences, then sets the co-association in motion to discover what results.

In this sense, a node is intimately involved in its experiment. It could even be said that, given it is testing itself, it *is* the experiment. The node has a vested interest in the experiment because it evolves according to how the experiment plays out. To the extent that the Dao eventually receives everything nodes experience, it could be said that the Dao equally has a vested interest in the result of each node's self-experimentations, as it also evolves according to the result.

Because physicality enables nodes to test and stretch themselves in ways they cannot in the purely spiritual domain of the Dao, almost all nodes choose to co-associate with bio-identities at some stage of their evolution. We won't say all do, because that is not the case. And there are varying degrees of involvement, which give rise to varying intensities of experience. There are also shorter or longer durations of co-association. All these parameters are selected by nodes themselves.

Despite the varieties of approaches available, for any node to successfully co-associate with a bio-identity one requirement is essential for all: each node needs to find a species that suitably matches its complexity.

Matches are identified in the same way that environmental niches suitable for nurturing biological species are found: nodes explore the multiverse and find them. However, this is not an inexperienced or naive exploration. Nodes have been exploring the multiverse for the entire duration that physical matter has existed. So accumulated information exists at the level of engendered Dao-consciousness as to which biological species are best suited for which type of node.

To present a somewhat tongue-in-cheek comparison, the process of identifying suitable species is like people deciding which country they will visit for their summer holiday. Many in Britain go to Spain and the Mediterranean. Those in Russia go to the Baltic. Those in North America go to the Caribbean, while those in Australia and New Zealand to the Pacific Islands and Asia. Why are favoured holiday haunts divided in this way? It is due to simple convenience. No one wants to spend a substantial period of their holiday time travelling. So they find places that have a suitable climate and are conveniently close.

Furthermore, there is plenty of quality advice available, from travel agents and friends, regarding what each potential holiday location offers. There is also your own knowledge, given you may have passed through a place, liked the look of it, and so decide to return for an extended holiday.

In the same way, when a node seeks a suitable species to co-associate with, advice is available. Mature nodes that are similar in Dao-nature will have previously explored certain species and found them suitable for co-habitation. They then recommend the same species to other nodes who are just starting out on their explorations. So it is (again we speak with our metaphorical tongue firmly in our metaphorical cheek) that advice is available to the node from the equivalent of travel agents, who have knowledge of certain places within the multiverse and who are able to make recommendations based on what is convenient and appropriate for the node.

No node is told where it should go. Each selects its own destination species. Just as the holidayer does, the node seeks advice from spiritual level friends, and also draws on its own perceptions and experiences, before selecting a species and deciding what type of co-association it will have. To bring this analogy to an end by showing the limits of its application, an extended sojourn co-habiting with a species like the human is in no way a holiday.

Identity is bi-located

To reiterate our central point, once a node or node fragment co-associates with a physical body it becomes a dual identity. It possesses Dao-consciousness and bio-consciousness. One way of describing this is to say a human being is a bi-located identity. It possesses consciousness in two locations. Simultaneously.

This last is a significant point. At any moment human consciousness and identity exists in two locations: in the physical realm and in the spiritual realm. This means that at any moment either, or both, can be accessed. The fact that they are not, that in fact many are convinced their identity is mono-located, and only in the physical, is clearly because they are not aware of their own bi-located nature. Becoming able to direct one's awareness on both levels is part of the process of maturing as a fragment of a node.

With bi-location clarified, next we will examine the factors that come into play during embodiment and facilitate the maturing process.

8

The Process of Embodiment

AN EMBODIED HUMAN BEING finds it very difficult to imagine the full range and richness of opportunities available to a node of Dao-consciousness. The node selects from those opportunities in order to achieve spiritual maturity.

Physical species available for co-habitation exist throughout the multiverse. There are innumerable galactic and planetary systems from which to choose. Some identities elect to specialise; some do not. The choices a node makes leads it sometimes to one species and sometimes to another. The experiences available across species differ in degree and kind. They also differ in the opportunities they present to explore either solitude or a social life.

For example, the modes of consciousness according to which an ant encounters another ant, and interacts with it within the social group of the colony, provides an opportunity to explore trust in the greater good. For ants trust consists of being willing to act as an individual yet unite with common purpose in the interests of the colony's survival. This involves acting fearlessly in the face of a predator and cooperatively carrying out other tasks. The activities of the colony engage the attention and loyalty of each ant. An ant mind is subservient in nature. Besides colony activities, little else distracts it. Hence each ant comes to understand subservience as a way of life. So while an ant's awareness is confined compared to human awareness, that awareness is sufficient to enable any ant to thoroughly explore quite specific categories of social life.

Why this planet for embodiment?

The rich range of available opportunities throughout the multiverse make this planet just one of innumerable possibilities for embodiment. It is simple convenience as much as anything else that leads a node of Dao-consciousness to occupy itself with the generation and extinction of karma on this planet, as opposed to exploring any opportunity available elsewhere. In that sense, there is nothing special about this planet. It possesses a variety of environmental niches, and in many instances it offers a relatively benign environment.

What primarily draws spiritual identities back to this particular planet is a need to extinguish karmic debts with those with whom it has generated them. In fact, this human animal species offers opportunities that are mostly mirrored in other species in other planetary and galactic systems. Occasionally one of them is chosen for embodiment as an alternative to coming back to this planet.

From the human perspective, there is an inevitable expectation that a node fragment will occupy a land species such as a human being rather than a member of its cousin species that live in the oceans. In fact, each species provides opportunities to undergo particular kinds of experiences.

However, a node fragment will generally not switch species. This is because the parameters of control tend to differ wildly between species. Acquiring the attributes needed to control the organism of one species is so challenging that only particularly courageous individuals attempt to learn to control multiple species. That is sufficient reason for choosing to occupy one particular species and to disregard others.

Embodiment is a pre-planned opportunity

Engaging with a physical life is a creative act. The creativity starts before an individual begins its co-association with a human embryo.

Co-association results from a spiritual identity sensing an opportunity to develop a useful new sub-identity, a new human personality. An individual's choice of a particular physical body and its associated psychological makeup usually follows a favourable calculation for matching the intentions of the node fragment with opportunities it will have while embodied. One of each node's prime purposes is to extricate itself from a variety of self-limitations that it perceives are a barrier to it fully flowering as an autonomous, independently mobile, well informed and mature individual product of emergence from the Dao.

Being able to appraise future possibilities comes naturally when individuals occupy the pure domain of Dao-consciousness. From that perspective embodiment can be seen to offer the opportunity to interact with old friends, to engage in various forms of mischief and amusement, and to act with serious purpose. Thus the individual engages with others with whom arrangements have been made to meet, to interact, and to work out karma together, or perhaps even to make karma together—although this last is not usually an objective.

Embodiment in the physical domain also offers rich opportunities to meet and associate with others quite unlike the incarnating individual. This, then, becomes another significant opportunity to meet others who have a radically different make-up, perceptions and attitudes, and to experience the arrogance, self-grandeur, slimy obsequiousness, even expressions of outright evil, that occur in the human domain. Loving individuals have no opportunity to experience these characteristics in their normal location in Dao-consciousness.

Pre-selecting cultures for individual growth

Optimally, an individual selects a body, develops it, and lives out a life that comes to a successful conclusion. Naturally, there is a range of possible outcomes, from the sub-optimal, to the optimal, to the ecstatic—by ecstatic we mean an individual's efforts may lead it to ecstatic states being experienced during that lifetime.

Where an individual grows up in a culture that provides optimal rather than sub-optimal conditions, opportunities are freely available to attend to various trauma and to deal with their emotional and psychological impact. The purpose is to clear them so they no longer influence the developing organism. The result is that the individual develops its capacity to process physical, emotional and spiritual input optimally rather than sub-optimally.

Where an individual grows up in a cultural environment in which supportive opportunities are not so obviously available, it simply makes its way as best it can. Within such a culture, if the individual, due to its own inner qualities, comes to experience intensely ecstatic or erotic states, it will likely be labelled as acting beyond the pale, being in the wrong, or being outright crazy. These encultured responses then become part of the experience of living that life.

Cultures are extremely complex in their organisation. It is possible to categorise entire civilisations, and sub-cultures within each civilisation, as providing environments that are inadequate, adequate or better than adequate in terms of supplying the conditions an individual requires to live an optimal life. However, it is also true that almost any civilisation contains a vast range of opportunities. Accordingly, individuals are usually able to find a place within the culture they are born into that accords with what they seek to achieve in that life. This implies that a monoculture imposed by an authoritarian ruler is inadequate. So is a culture subservient to another external culture. What is adequate is a condition whereby an individual is free to sample every input available in the surrounding culture and so have the experiences it planned prior to incarnation.

During the pre-planning phase each individual is given information on the range of options available for its upcoming life. The individual is informed that the range is extensive, and that it has complete choice. Choosing includes being able to request further information should the individual elect to enquire more deeply into what is available. No blame is attached whatever it chooses. All experiences become part of its ongoing education.

In this way each individual's placement into a culture is facilitated so that it accords with the individual's life plan. The individual is free to pre-select sets of individuals to interact with. It also chooses from a range of possible relationships it may have with each. And while living the life it is free to make choices in relation to each of those significant individuals.

Where the outcomes of relationships are positive, by the end of its life the individual will have come to perceive that it had a purpose in associating with humanity, and that humanity in turn supported its development and enquiries. Optimally, by the time a life ends it will have been able to experience those activities it had chosen prior to incarnation, and it will feel pleased because it carried out those activities in a way that satisfied its life plan.

For the ordinary individual growing up in a modern culture there are endless opportunities for making optimal choices and sub-optimal choices. No choice is wrong. But all choices bring consequences. It is through facing up to consequences, and particularly facing up to the impact those consequences have on others, that rich karma is generated. Or, alternatively, karma is reduced.

In a free society, all of these activities constitute grist for the mill of the node fragment as it manifests its intentions. And properly so.

Embodiment issues for naive nodes

Individuals who are naive, that is, who are yet in the early stages of their experiential development as a unit of Dao-consciousness, may come into the human realm and make decisions, but be ill-prepared to endure the consequences. Given that during the early stages of their development they are still unfamiliar with the controls needed to occupy a human animal body, when they make decisions they are likely to be over-ridden by human animal biological impulses. For the naive, their attempts to control the rampant human emotions they feel overtaking them will likely be ineffective.

Human emotions are the product of a very long period of competitive socialising with both enemies and peers. In addition, over the eons humans have been prey for several other species. So normal mammalian emotional and biological reactions experienced during any life often give rise to opportunities for provocation and response—such as manifesting fear, avarice, power or anger—which consequently generate considerable karma. Karma is generated when others are radically disempowered, confined or killed. Reactions that generate karma often lead to consequences within that life itself, including retribution at the individual, familial or clan level. These chains of reactions generate potent situations and feelings that need to be resolved in later lives. It is necessary first to engage in restorative action, then to eliminate each and every instance of karma-generating impulses and karmic indebtedness.

There is much to be learned in all this activity. In particular, individuals progressively develop an awareness of the contrast between reactions involving strong human emotions, including anger, provocation and jealousy, and feelings emanating from the intrinsically loving nature of a spiritual identity and of the Dao-consciousness that comprises it.

Frequently a node fragment is appalled when it appraises the results of what proved to be a naive wish to engage with others in the human domain. The individual then usually retires to reflect at length, seeking to discover and understand what motivated its reactions and choices. It then undertakes to repair all it did during the course of that single human life.

The distinction we are making here is that, over the course of multiple incarnations, there are parameters of control. There are also various measures in place that individuals may use to progressively understand the ways a life form may be modified for the better. This includes appreciating which sets of human animal reactions can be inhibited and which cannot, and ceasing to engage those that contribute to the individual's ongoing limitation.

The nature of masterful nodes

Continuing this discussion of what is typically experienced during a full cycle of incarnations, we will now consider the masterful fragment of a node of Dao-consciousness—we deliberately express it this way to avoid our description being contaminated by traditional terminologies.

Masterful node fragments thoroughly understand the characteristics inherent in the body with which they are co-associating. They are able to influence specific aspects of its personality development. As a result of wisdom accumulated over multiple lives, the masterful node fragment chooses certain sets of individuals to mix with, avoiding others. The knowledge mature spiritual identities exhibit is reflected in their movement towards the positive and away from the gruelling.

Of course, what the masterful node actually does is lift the horizon of its regard from the mundane, superficial and inconsequential, and instead seek opportunities to benignly and constructively influence the society it inhabits. This facilitates breadth of perspective and depth of understanding and leads others to view it as possessing wisdom and maturity.

These constitute the varieties of opportunities created when an individual node of Dao-consciousness engages in human embodiment. It is for personal enrichment, which becomes transpersonal enrichment when all it experiences, learns and comes to understand is uploaded to the group soul of which it is a small but vital contributing aspect.

Developing subtle grace

When an embodied identity becomes aware that it not only consists of its current body and personality, but that it also possesses a series of prior personalities accumulated at the level of the spiritual self, it can be said to know itself fully for the first time. Of course, the spiritual self is usually only perceived faintly and fleetingly. That is because fulfilling roles and duties, and sustaining the physical focus needed to survive, inhibits anything beyond a superficial knowledge of the spiritual self. Nonetheless, it is enough for an individual to have confidence that this is the structure of its being.

Re-establishing a connection to the accumulated self at the spiritual level usually requires that the individual achieves a state of enlightenment. For some that appreciation only occurs when they do not need to incarnate any more. Knowing other times and places as intimately as one knows the present time and place inevitably leads to an individual living with subtle grace. Subtle grace is one of the distinguishing features of an experienced individual. It is a product of having lived many lives.

The term subtle grace may be applied both at the energetic level and to the way an individual conducts themselves on their personality level. It results when an individual confidently knows what their identity is composed of. Of course, their identity is multiple. Such an individual recognises more than just the present time and place. It knows it has repeatedly engaged with the planet and its populations, and it understands that it always survives, despite the body's death. This graceful individual is free to grow its understanding without feeling a need to endlessly question and challenge the prevailing view, whatever that might be. Such an individual has the capacity to act simply and effectively, skillfully negotiating the challenges and problems of life.

Such individuals are commonly found among those who assume responsibility at the societal level. Of course, there are many other types of ambitious individuals who strive to take their place at that level, but who demand and command without possessing the qualities of knowledge and subtle grace we are referring to.

The term *subtle* relates to the sense of self-command achieved by the mature node fragment who has experienced many hundreds of lives and innately understands human existence. It is able to instantaneously analyse every situation it encounters. Its self-command is reflected intellectually in calmness of mind and emotionally in its unperturbed mien. On the energetic level this individual is a centre of peace rather than of turbulence.

These qualities lead others to pick out such individuals as preferred figures of authority. They commonly accept management responsibility and positions in which they control others. Frequently, this is in the realms of business, the professions and the military—although by that phase of development conflict as a means of problem resolution is often already abhorred.

Collectively, the qualities associated with subtle grace involves the capacity to perceive, the capacity to analyse, the capacity to understand, and the capacity to command. That combination is commonly interpreted as representing intelligence. But it is more than that. This is why we have taken this moment to clarify the distinction.

Those who desire to learn more about that class of individual and their distribution through every culture, or who seek more information regarding their multi-life trajectory, may find it in the Michael Teachings. That set of descriptions offers enquiring individuals a more detailed understanding of their place in the spectrum of embodied identities. However, bear in mind that all such analytic and descriptive systems are merely intended to provide a broad background understanding of the conditions of life and love. In that sense, those writings, like this text, are just another attempt to bring order to human understanding of the self.

9

How a Spiritual Identity
Connects with a Body

WE NOW NEED TO CLARIFY the actual process by which a node co-associates with a bio-identity. In what follows we will principally refer to what happens when the fragment of a node of Dao-consciousness co-associates with a human body. This is purely pragmatic, because we are addressing those occupying the human domain. We begin at the embryonic stage.

How a node fragment merges with an embryo

A node fragment of Dao-consciousness, being an individual spiritual identity, first associates with a human body when the body is an embryo in its mother's womb. A particular embryo is chosen for the reasons outlined in the previous chapter. An intention to co-associate is then directed towards the embryo. Beyond this there are no particular levers to pull or any kinds of biological or energetic structures to manipulate.

Given that a physical body, like all matter, mostly consists of space, and given that the spiritual identity does not occupy space, the dimensions of co-association are not physical. Simple intention is sufficient to ensure the spiritual identity and its chosen body are united for the duration of that body's life.

The blending phase requires the insertion of a point of attention. A node fragment has only to focus its intention, then sustain that intention, for there to be effective co-association between the spiritual Dao-identity and its selected physical bio-identity.

The process is similar to the way that two particles of fog merge. The ionic boundaries of one droplet blends into the ionic boundaries of the other droplet to form a slightly enlarged ionic boundary comprising both sets of molecules. Similarly, it may be envisaged that a "droplet" of organic matter merges with the "droplet" of spiritual matter. Even though the material of each are different, they intersperse with each other and form a composite whole.

This is all that is required. That is why it is called merging, coalescence or co-association. One merges into the other and a new composite identity, a bi-located identity comprised of Dao-consciousness and bio-consciousness, comes into existence. This occurs while the foetus is still in its mother's womb.

Of course, more is involved than just the directing of an intention. We will now discuss in more technical detail how this merging is able to be sustained.

The globular structure of a spiritual identity

The nature of the body, including its basic shape and functioning, is well known. What is not well known is the nature and shape of the spiritual identity that locates itself beside, within and through the body. The individual spiritual identity, and we are talking now in an energetic sense, exists in the shape of a sphere. Its structure is globular.

Imagine a spherical ball, say a soccer ball, with the hexagons and pentagons forming a net around its surface. Now imagine every intersection on that ball connected radially both inwards and outwards, forming a three-dimensional structure consisting of cellular interconnecting filaments. Here the image of the hexagonal walls of honeycomb inside a beehive are appropriate. The cellular structure inside the spirit sphere is patterned like the beehive, but much less rigidly, certainly not hexagonally, and the filaments extend inwards and outwards throughout the entire globular structure. The interconnecting filaments provide the means by which encoded information is carried through the entire structure.

This spiritual structure is not fixed in any particular place. However, it does assume a level according to its attributes on what we term agapéic frequency. We will discuss this more fully in the chapters dedicated to agapéic space. Here we will introduce the concept by saying that any individual spiritual identity is situated in agapéic space according to its agapéic frequency, hierarchy and willingness to bequest agapé. Because most human individuals are at similar but slightly different values on those axes, they exist at adjacent positions in agapéic space. Differences are naturally also present in the information encoded within each individual globular spirit sphere.

Hence those who are at similar levels of experience, development and maturity may be imagined as forming clusters of globular structures existing adjacent to one another in agapéic space. This is a crude approximation, because none in agapéic space feel they are crammed on top of or merged into any other spiritual identity. But it gives a reasonable introductory albeit abstract approximation of what is the case.

How information is transferred between levels

Encoded within the spiritual identity's globular structure is information regarding what has been done in previous lives and what is intended to be achieved in the next upcoming life. The question then arises, how is this information transferred from the individual at the spiritual level to the embryonic human individual at the physical level? How is information regarding the life plan communicated by the spiritual self to its animal human self? This type of information is transferred from the spiritual identity to the human animal identity via the aura. The aura in turn exists on the electrospiritual level. As these structures are described at length in the next chapters we won't discuss them here.

So when a spiritual identity despatches an aspect of itself into the human realm, the now born and growing human being can access knowledge of its spiritual self's goal in this life, along with what it has learned from prior lives, and all other information encoded in its globular spiritual structure, via the aura.

When the body dies and is incinerated or decays, the aura itself dissolves. But in the process of dissolving, all the information contained within the mind of the individual's human physical self is spontaneously uploaded, to use that modern concept, as patterns of information conveyed to the globular spiritual identity through the connecting link of the aura. This is easily achieved, because the identity is bi-located, with its spiritual self and its bodily self nested beside and inside one another. There is no distance for the information to travel. The uploading is direct, spontaneous and instantaneous.

How the spiritual and physical are aligned

The final point we wish to make here is in relation to how the globular spiritual identity and its animal identity come into alignment and sustain that alignment for the duration of a lifetime. We mentioned that all that is required to establish a connection is for the spiritual identity to form an intention to co-associate and it is is done. However, there is a central point, so to speak, at which co-association is achieved.

Within the globular spiritual structure, at its centre, is what might be thought of as a kernel. This is the centre from which all the filaments radiate. It could be conceived of as being somewhat like a small seed within a translucent globe covered and filled with patterned filaments of light. But, of course, the kernel exists energetically, not physically.

When a spiritual identity seeks to co-associate with a body, it brings the centre of its globular structure into alignment with a central energetic point within the body's aura. This energetic point becomes the hara dantian, just below the belly button. In this way the centre of the spiritual globular structure is aligned with the centre of the physical body. Intention then sustains this alignment for the duration of the life. The alignment with the aura's energy system also facilitates auric communication between the mind of the spiritual self and the mind of the bodily human self.

We previously mentioned that everything requires preparation. The inexperienced node fragment needs to learn how to align its kernel centre with the aura's hara centre in order to sustain its connection with a physical body. Only after learning to do this can it begin co-associating with the body of any species.

This gives rise to an image, which has appeared in spiritual literature through the ages, of the body surrounded by the slightly larger envelope of the energetic aura, and the aura surrounded by the even larger globular structure of the spiritual identity, which is actually somewhat oval in shape, not strictly spherical.

To conclude this description, and to make what we have said clear, no binding or locking into position is required for the spiritual self and the human bodily self to be aligned. When there is an intention to merge, merging occurs. When there is intention to dissociate, merging is terminated. The identity, which is always bi-located, has complete freedom to leave its association with the body for a time, which is naturally done during some phases of sleep, or to sever the connection completely—even, if it chooses and for whatever reason, it chooses to do so before the natural death of the body.

Bi-location involves dual sets of information

A significant consequence arises out of bi-location. This is that the ordinary mind of each human being has two quite separate sets of information coming into it. One set comes from the spiritual self, which urges its current human identity to act in accordance with the plans it has made for this lifetime. The other set comes from the body, its hormonal system, and the human identity's socially conditioned behaviours, feelings and thoughts.

The first set of information is subtle and delivered quietly into the ordinary mind. The second set is coarse, loud and insistent, and quite naturally engulfs the individual's attention. Of course, the spiritual identity has chosen to enter the human body and its complex world to experience what is on offer. So the fact that its subtle intent is drowned out by coarse body level input is not just understandable, it is necessarily a part of the challenge all embrace when they take on the difficult task of learning by engaging with the human.

As we stated earlier, learning to become a balanced bi-located identity, in which Dao-identity openly and appropriately contributes to life experience alongside bio-identity, is an essential aspect of what is involved when a node fragment utilises human experience to achieve maturity.

The Aura and
the Electrospiritual

10

The Electrospiritual and
the Implicate Order

IN THE PREVIOUS CHAPTER we described how the globular spiritual structure centres on the hara dantian location in the aura. This facilitates the alignment of an individual Dao-identity with a bio-identity. The aura facilitates the flow of information between the two levels of bi-located identity. In this chapter we will provide further detail on the interconnecting structure, given the aura is actually just one aspect of what is an extended aspect of reality.

But before we begin this explanation it is necessary to remind you, the reader of this missive, that the way we intend to describe these matters will make it natural for you to infer a hierarchical relationship exists between levels. We wish to dispel this notion. The model is one of interpenetration. Just as, for the individual, the globular structure of the spiritual identity interpenetrates the body's flesh, bones and nerves, so at the level of the kosmos the spiritual domain interpenetrates the physical domain. Similarly, the interconnecting structure we will now discuss interpenetrates and is interpenetrated. All interpenetrates. Nothing is separated into the kind of hierarchical relationship that comes naturally to human thinking. Certainly, different structures exist on separate levels or frequencies. And one level is not necessarily perceived from another. But they are all equal. None is more important than the other. Collectively, they all constitute the extensiveness of what is.

Having clarified this, we wish now to describe a particular feature of the structure that connects the spiritual with the physical. We call this interconnecting structure the electrospiritual. In order to put it into context we need to return to a consideration of how the multiverse came into existence.

The Dao intended the electrospiritual

No doubt some enquiring readers of this transmission will be wondering how the physical universe came into existence. In order to discuss the nature of the electrospiritual we need to address this question.

Just as the Dao casts nodes from itself, so the Dao also cast out of itself all the universes that constitute the multiverse. Science postulates the existence of singularities, and that this universe humanity lives within came into existence via a singularity. Accepting this notion, the question is, how did the singularity come into existence?

In the vision of the multiverse we projected into the mind of Peter Calvert, the model showed connecting filaments or threads stretched between the separate universes and the *observer*. We make two further comments now regarding that model.

The first is it implies a hierarchical relationship, with the *observer* some distance from the universes. As we just noted, this must be translated into an interpenetrated model, in which the *observer*, the connecting threads, and the universes, interpenetrate each other. Ken Wilber's notion of holons is appropriate here, with reality consisting of a number of nested layers that are simultaneously individual parts and inextricably parts of the whole.

The second point is that the threads represent the electrospiritual. We identify three layers as basic to the kosmos: the electrospiritual, the electromagnetic, and the electrophysical. The electrophysical refers to physical bodies constituted of chemicals and electrical impulses within the nervous system. The electromagnetic, constituted of microwaves, x-rays and the like, is well known. The electrospiritual is postulated as an extension of these frequency fields. Just as the electromagnetic consists of finer frequencies than the electrophysical, so the electrospiritual consists of finer frequencies than the electromagnetic.

When the Dao gathered an intention within itself to manifest the multiverse, an intention we have characterised as the *observer*, what happened can be likened to a fluctuation. That fluctuation manifested the electrospiritual. Within the electrospiritual were encoded patterns of information. That information spontaneously formed a singularity that resulted in the coming into existence of this universe.

We are again trying to describe what cannot be described. The danger is that the human mind automatically reduces any concept to human size, domesticating what is actually inconceivable. We are presenting a model here, sketching a bite-sized picture, using current ideas that relate to scientific conceptualising. So our description may best be considered a vague approximation. It is a work-in-progress. No more.

How the Dao created the multiverse

The creative act of something coming to be out of nothing, or at least out of no other material substance, has long puzzled humanity's best minds. In order to understand what creation is it is necessary to extend one's conceptualising beyond the strictly electrophysical and embrace the notion of the electrospiritual. This will be anathema to many, but that is beyond our ability to change. However, given reality consists of the spiritual as well as the physical, we make no apologies for our stance.

Elsewhere we have used the metaphor of condensation to indicate in concrete terms what we are trying to convey here. Water molecules are normally suspended in the air as vapour. The vapour's molecules are too small to be visible to the human eye. But as the temperature falls, the molecules of water vapour come together in a process of condensation. Eventually, they become sufficiently large to be visible to the human eye as fog. By the process of condensation what was invisible vapour becomes visible fog. In a similar sense, by a process of kosmic condensation a universe that did not previously exist came to be.

Where likening condensation to creation fails is that the water molecules were already present physically, they just weren't visible. In contrast, before the universe was created nothing was there at all. In this sense condensation is unlike the act of creation. Nonetheless, the universe's creation via a singularity can be likened to spontaneous accretion or condensation, being of course from the spiritual realm to the physical realm, rather than physical to physical.

Another non-everyday example of the type of creative process involved is provided by the zero-point field. Quantum physics proposes that matter and anti-matter spontaneously arise, cancel each other out, and vanish again. This field of virtual particles constitutes the zero-point field. However, when symmetry is broken the field becomes a non-zero field, and a non-zero energy state results. Matter then exists. In these terms, the universe itself may be postulated as being a non-zero energy field. The process can be described utilising the Laplace transform, as has been done. More simply, matter may be thought of as condensate from the zero-point field of symmetry-breaking states within the electrophysical. Of course, this does not apply to the already existing universe coming into existence, nor does it explain the occurrence of the originating singularity. But it does suggest another way that one zone of reality may translate into another. Because, to remind you, what we are attempting to describe here is a process by which the physical was created out of the spiritual.

That done, we will now bring this somewhat abstract discussion back to the human level and consider the nature of the electrospiritual.

The implicate order

The physicist David Bohm has named a subtle component of reality the implicate order. We here adopt his term and use it to refer to the patterns of information embedded within the electrospiritual.

Every living creature's body, whether it be an amoeba, a fungus, a plant, an insect, a crustacean, or an animal, is a configuration of the electrophysical into a distinctive biological form. Each species' distinctive bodily form, and hence the bodily form of every individual within each species, is a result of patterning that comes from the implicate order. That is, each body's existence is facilitated by the ordering of its electrophysical structure from the level of the electrospiritual. This somewhat opaque statement needs clarification.

The human body, to use it as an example, begins its life as a zygote formed by an egg and the sperm that fertilises it. The zygote's cells rapidly multiply and the developing foetus grows. Some cells become flesh, others nerves, some form the backbone, yet others the kidneys, heart, lungs and brain. This process involves undifferentiated cells turning into specialised cells that graduate to specific places on the endoskeleton and perform particular functions within the growing body.

DNA conveys traits to the developing body at the genetic level. But we maintain that the overall body is structured not entirely by DNA but that significant information is conveyed from the electrospiritual implicate order that tells the growing foetus' cells where to go and what to become. In effect, the cells conform to a higher level ordering dictated at the implicate level.

Clearly, this is a contentious statement. It will take considerable research to confirm. Eventually it will be done. At this stage what we have asserted can only exist as an unconfirmed proposition—alongside much else proposed in this expansive text.

In summary: The electrospiritual provides patterning to each and every living being in the electrophysical domain. The implicate order within the electrospiritual contains specific blueprints, according to which the cells that constitute the bodily makeup of each species conform.

Within these blueprints is the intent of the *observer*. It is too simplistic to say that the *observer* designed each blueprint in the implicate order. It did not, and we are not saying so. A much more layered, subtle and creative process is involved in ordering at the implicate level, which involves the intent of nodes of Dao-consciousness.

We will describe this process in the following chapter. But before doing so we will draw together the threads of the model we have proposed.

A new model of the kosmos

When viewing this model it is necessary to remember that, as the human saying goes, the map is not the territory. A two dimensional map is always an abstract approximation to three dimensional terrain. In this case, a two dimensional graphic cannot convey the layered multi-dimensional complexity of the richly experiencing Dao. Nonetheless, this is presented because it has been asked for.

There is the Dao, the ultimate unmanifest, from which everything ultimately derives. Within the Dao is the *observer*, which is a direction, an urge, a tidal swell, that is the Dao's intent to create. [See FIGURE 2.]

Out of this intent the Dao manifested the electrospiritual, the implicate order within the electrospiritual, and the electrophysical. Patterning across all three connects the manifest physical multiverse and the creatures and structures within it with the initiating wholly spiritual creative intent of the Dao.

Human beings may come to an understanding of these different aspects of reality because each individual human being is constituted of the electrophysical, the implicate order in the form of the aura, the electrospiritual, and the creative intent of the *observer*. Ultimately, all exists in, of and from the Dao.

As a final reminder, this model is presented in a vertical way that suggests a hierarchy. To this degree it is incorrect. All layers are nested within the other layers, as holons. It could be thought that the *observer* is in the centre and the manifest physical domain is on the outside. Or the reverse could be thought. In fact, there is no inside or outside. There is only an integrated whole.

A NEW MODEL OF THE KOSMOS

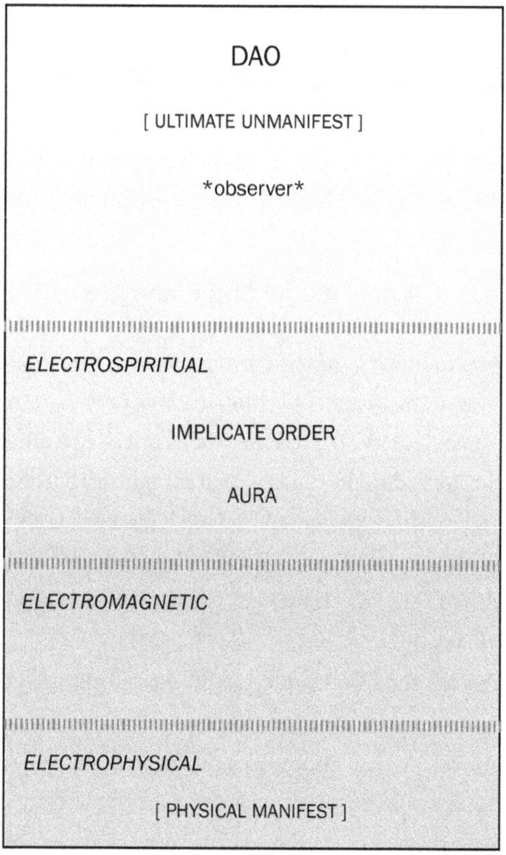

DAO

[ULTIMATE UNMANIFEST]

observer

ELECTROSPIRITUAL

IMPLICATE ORDER

AURA

ELECTROMAGNETIC

ELECTROPHYSICAL

[PHYSICAL MANIFEST]

FIGURE 2

11

The Nature and Purpose
of the Aura

WE IDENTIFY THE AURA with Bohm's implicate order, as explained in the previous chapter. In the sense we are discussing here, the aura is a structure within the implicate order.

This means the aura is electrospiritual in nature. It is a function of the implicate order. One of its key functions is to contribute to the body's form and natural qualities. In the previous chapter we asserted that the implicate order provides blueprints that shape the general patterns present in the electrospiritual into specific species patterns. The aura facilitates this process.

How the aura shapes organic life

The aura's emergence from the implicate order begins when the sperm first fertilises the egg and the two fuse to form a zygote. The aura can be thought of as a local accretion of the electrospiritual that condenses around the foetus. This is not unique to human bodies. Every body of every single living creature, including plants, insects and animals, possesses an aura.

For each species, specific patterning relevant to a creature of that species manifests out of the implicate order. The aura's function is to channel that information to the growing body's cells. In effect, the aura is the means by which the plastic undifferentiated cells are imprinted with instructions from the implicate order regarding their particular location on the endoskeleton, and fit themselves to their appointed biological function—bone, flesh, brain tissue, and so on.

Obviously, the genes also contain information relevant to the development and formation of a specific body. These relate to details: hair colour, eye colour, propensities to certain abilities and illnesses. In contrast, the aura transfers broad species information to the foetus.

Therefore the aura may be described as performing a dynamic organising function for the construction and maintenance of an organic body. At the end of a body's lifetime, when the need to maintain that physical organism ceases, the aura returns to that from which it came, the electrospiritual.

Why are we making so much of the electrospiritual in relation to foetal development? And how is it that the implicate order is so significant in particular to humanity? The answer is that the human aura was shaped in the distant past by Dao-identities whose aim was to bring the human form to its optimal condition so node fragments could co-associate with them and partake of the most complex experiences possible within the human domain.

We note that this type of auric level modification, even of interference so to speak, no longer takes place. This means that the association of the growing aura within the growing body is now an automatic process. The aura is imprinted by the implicate order, and the aura automatically imprints the growing cells.

How personal qualities are imparted to the foetus

Before incarnation, a node fragment not only considers where it wants to live, which culture it will exist within, and what family and therefore body it will inhabit, it also selects certain specific character traits and abilities. Some of these will likely already exist in the genetic makeup of the selected foetus, being factors that contributed to that body being a suitable option. But other traits, not supported genetically, will likely also need to be transported to the new identity.

This process could be termed fine-tuning, in the sense that it involves subtle information regarding character traits—predilections, what human beings call natural abilities, and so on—being embedded in the nascent identity. This information is embedded at the auric level.

So the aura contains two levels of information. One is impersonal, being general information regarding the shaping of the cells into a human body. This information comes from the implicate order and is automatically encoded into the aura. The second level is personal. This is information that the node fragment deliberately imprints itself into the aura. It consists of traits and abilities selected to facilitate specific kinds of life experiences.

How characteristics are carried from life to life

For a spiritual identity to co-associate with a bio-identity, all that is required is the intention to do so. Similarly, for a spiritual identity to influence the nature of its selected human body and personality, all it needs to do is focus its intention. Of course, this is a learned activity. So as an identity experiences more and more of the human domain, and as it develops greater skills at functioning in an aware manner, by which we mean a spiritually aware manner while embodied, so the identity also becomes more skilled at directing the development of its embodied selves at the auric level.

Dao-consciousness and the electrospiritual possess the same basic nature. As a consequence, the implicate order may be altered when a trained node fragment directs its intention at it. The implicate order is sufficiently mutable that it allows specific characteristics, qualities, malformations, or whatever else is intended by the node fragment, to be imprinted into the physical body via the aura.

This does not mean that a node fragment is able to transform its newly selected bio-identity into a great sports person or musician. It cannot transfer traits that it has not already developed in prior lives. What it can and does transfer are traits it has developed through its own efforts. It imprints these into its new body's aura and so influences the development of its new identity. This occurs because a node fragment recognises that certain qualities will optimise the physical and psychological nature of its new incarnation.

All this occurs at a level of which any currently embodied individual is unconscious. However, everyone can learn what qualities they have brought into this life, and why. They access this information via the aura.

The aura, chakras and nadis

To complete this description of the aura, we will now introduce further details regarding the energetic relationship between the body, the aura, and the electrospiritual. We begin by distinguishing between the components existing at the haric level within the energy pattern of the human body. What do we mean by haric level? We have previously described the kernel at the centre of the spirit sphere. When the spirit sphere co-associates with a human body, the point of alignment is at what the Japanese term the hara centre, just below the belly button. Accordingly, we identify the kernel as the hara dantian. [Japanese *hara*, belly, Chinese *tan tien*, energy centre.] So the haric level refers to what is experienced by the spirit sphere when co-associated with a human body.

To further clarify this terminology, when a body dies the spirit sphere's co-association with that particular body ends, along with its co-association with the body's hara centre. Hence the terms hara dantian and haric level do not apply when an individual is in a non-embodied state. An argument could be made that all references to the term hara dantian should be deleted, as it is entirely a product of body-centred awareness and does not reflect the true nature of the spirit sphere as a separate construction. Nevertheless, we will continue to utilise this terminology as long as required.

In addition to the terms hara dantian and haric level, we also introduce the traditional terms of ida, pingala and shushumna. The energetic nature of human reality is more complex than these terms suggest. However, we are offering a simplified description of the situation so embodied individuals may understand something of what is involved. We also invoke these ancient terms in order to bring long utilised sets of understanding into alignment with our modern rendition. So ours is an inclusive intention, not an exclusive one. We repeat, the model we are about to explain is a selected subset, not a complete description. This suits our purpose, which is to provide a means by which energetically structured reality may be understood in general terms by those who are incarnate.

The ida, pingala and shushumna are traditionally used to describe major nadis. [Sanskrit *nadi*, energy channel.] In our terms, the ida and pingala are inner auric-level flows. It is historically understood that the ida, pingala and shushumna are associated with the dantian. This is false. Furthermore, there are actually three separate levels, not two. We will explain.

An energetic bridging structure exists that connects these various levels with the body. The structure is significant because it directly contributes to the formation of the human body. As we have stated, from the moment when the egg is fertilised by the sperm, and a zygote consequently starts to develop, an aura simultaneously accretes around the foetus. It can be said that the physical body both generates the aura and is generated by the aura, in a reciprocal and mutually affecting construction.

Traditional spiritual teachings maintain that the aura contains chakras [Sanskrit *cakra*, wheel or circle], commonly known as energy centres. We contend that one of the purposes of the chakras is help guide the formation of the growing foetus. In turn, the chakras develop from the ida and pingala. And the ida and pingala are generated by the shushumna. The result is that the co-associating ida and pingala are matched and predetermined from a different level by the shushumna. So the chakras, the ida and pingala, and the shushumna, exist at three different levels within the electrospiritual. All naturally and spontaneously spring into existence, on their separate levels, when the parents' intention is concretised in the fertilised ovum.

Sensitive people have registered the presence of, or have independently perceived, the ida, pingala and shushumna in a relatively mature human animal. Recognising their causal relationship and association with the chakras, they have assumed that the chakras and ida, pingala and shushumna are either on one or two levels. This misperception has occurred because throughout human history practically no one has intuited that there is actually a three-layer stacking, if that correlation can be imagined. Neither has anyone observed the zygote energetically and so become aware of the multiple levels contributing to its development during the first trimester. Accordingly, for this fresh start, we state these things accurately.

To be clear, two independent constructions are being discussed. One is the manifestation of the fragment of a node of Dao-consciousness in its form as a spirit sphere and central kernel. The other is the developing zygote and foetus of the physical form on the level of the implicate order. Progressively, this involves first the shushumna causing the energy flows comprising the ida and pingala, the ida and pingala causing the progressive formation of the chakras associated with the developing biological construction, and the chakras guiding the body into its eventual fully developed form.

It may be thought that these two independent constructions overlap. That is, it might be thought that the process of formation we have just described is affected from the spiritual level, and particularly by the spirit sphere that intends to co-associate with the developing foetus. This is incorrect. The physical construction assembles according to the blueprint that is generated as a consequence of the parent's initial intention to create. Once a sperm fertilises an ovum, and zygotic development begins, cell fusion and multiplication naturally results in the rapid and familiar development of the embryo in accordance with its design within the implicate order. This applies to all species. Only when the human foetus is near term does the spirit sphere then begin its co-association, by intending the alignment of its dantian with the body's hara centre.

So the distinction can now be clearly made between the central kernel of the spirit sphere, its spherical or ovoid structure, and the pre-existing and independently developed causal principle, on the level of the implicate order, which leads to the formation of the physical body along with its energetic concomitant, consisting of the shushumna, ida, pingala, chakras and physical form.

Accordingly, there is no receptacle within the implicate order associated with the developing physical form that the spirit sphere "slots into." Rather, the centring naturally arises from the spirit sphere's intention to co-associate. The hara centre is simply the natural place by which the spirit sphere and body become proximate. That being the case, there is also a small subset of degrees of displacement that can generate malformation and related symptoms. These occur as a simple consequence of what could be termed a locational error between the spirit sphere and the physical structure, occurring on the three levels. Or, four, if the physical form is also counted.

So we have: level one, consisting of the sushumna; level two of ida and pingala; level three, consisting of the aura and chakras; and level four of the zygote, which subsequently develops into an independent body and its associated bioidentity and lower mind.

Verifying this model

We realise that, generally speaking, this model cannot be confirmed. Our formulation may also lead to unnecessary curiosity concerning the distinctions we have made between these aspects of reality.

Nonetheless, the single most significant experiential confirmation by this individual through whom we speak [Peter] was his personal observation of the ida and pingala in his Vipassana assistant teacher. Skilled practitioners may observe these classes of structure by taking their point of attention outside their physical form. From that perspective ordinary physical space becomes translucent, being perceived in a manner commonly described by people detached from their physical form, by which position it becomes possible to see and even move through physical objects. This class of perception has been reported frequently throughout history. It is also a feature of the out-of-body experience.

It is rather more unusual to perceive the shushumna as the causal and coordinating principle for the set of structures existing at the level of the implicate order associated with a living person. Nevertheless, there are compelling international databases of experience, to refer to them in that manner, in which individuals have attempted to accurately record their perceptions of such phenomena. Historically, these accounts have been encoded into symbolic forms. One such form is the caduceus.

The Western caduceus

The caduceus is a well-known symbol that was originally the rod of the Greek god, Hermes. Less commonly understood is that the cadeceus represents direct haric level perception of the energetic structure connoting the interaction between the spirit sphere identity and the aura level energetics of the physical body.

Considering the linkage between the caduceus symbol and the ida, pingala and sushumna [see FIGURE 3], the caduceus presents an image resulting from ancient perceptions of haric level currents within the structure of the organism. Having personally perceived the caduceus-like interior energetics of the embodied human, we are comfortable stating that this level exists. We also know that such a perception is not beyond any reader's potential comprehension.

Acknowledging this symbol is important for several reasons. It links perceptions across cultures and time, which is required of any description that seeks to be complete. It provides an opportunity to directly connect the obscure constructs, at least to Western eyes, of ancient Eastern perceptions with ancient Western perceptions. And it contributes to an appreciation that these patterns of energetic function are applicable to those living in all cultures and across all eras, up to and including the present.

On the other hand, it is also true that historical associations have become somewhat confused over time. This occurs because only rare individuals have perceived on the haric level and generated durable records of what they saw. Such symbols subsequently take on a historical life of their own, due to repetitious usage and people assigning them, at their convenience, from one sphere of activity to another, such as when the caduceus was confused with the rod of Asclepius and adopted by some in the medical profession.

Nonetheless, we maintain the caduceus symbol remains a valid representation from the mythic pasts of both East and West. We here make a formal link to this and other related symbols and terminology, in order to identify their deep origin and to affirm that they represent the existence of a haric level structure, as has been observed by seers throughout the ages.

THE CADUCEUS

The ida and pingala shown twined around the central sushumna, causing the interlocking pattern commonly termed the caduceus.

FIGURE 3

Haric level perceptions and beyond

As we observed, only a very small number of people have directed their attention in order to perceive on the haric level, well under one in one thousand. Individuals possessing the requisite intellectual mindset, determined to observe, record and think about such things in terms of models and the philosophy and psychology of perception, are rare within any population.

Given this is the case, the lone reporter can only claim perceptions as legitimate by referring to similar perceptions from other times and places. Fortunately, there is today no difficulty in finding such historically-centred discussions, as they are available on various websites.

We would point out, however, that there are good reasons for disagreements between older models and this one. It is because we make no mention of the higher and lower dantian, which are a construct of perceptions from yet other levels, which we will not comment on in any detail here. We will say only that interaction between the spirit sphere kernel and the shushumna, ida and pingala generates a resonance which produces the second and third dantian. A pattern on one level produces, through the phenomenon of sympathetic resonance, a pattern on a higher octave. No one currently alive has seen this. The memes propagated through history, and now archived on the internet, are derived from original observations now almost lost and inaccessible to those in the English-speaking world. Hence the other dantian are of theoretical interest only and will not be referred to again.

In stating this, we signal that perceptions do not end at the haric level. But this model is limited to the first dantian, which we have identified with the kernel of the spirit sphere. We limit this discussion because our aim is to provide a simplified and so easily understood model, which is then available to be promulgated into the world community. Accordingly, we do not wish to get lost in abstruse detail—and this is already sufficiently abstruse from the perspective of almost everyone.

Next we will consider the aura's communicative function.

12

Using the Aura to Communicate

BESIDES PATTERNING PHYSICAL CELLS the aura also acts as a natural channel of communication between a node or node fragment of Dao-consciousness and the bio-identity. We have just discussed how, when the foetus is growing, the aura transmits information from the implicate order to the growing cells that contains instructions regarding how and where to grow. In this sense, the aura acts passively, as a conduit.

This passive quality also makes it ideal for deliberate and consciously directed communications from the Dao-identity level. Furthermore, because the aura is a passive conduit, communication easily goes both ways. This means that when an individual functioning at the level of its ordinary mind wishes to communicate with its Dao-identity (however that identity might be conceived, and people have many different ideas regarding it) the aura facilitates the flow of that communication.

This function of the aura is seldom appreciated. Often when people experience communication from a node of Dao-consciousness they think something divine has touched them in some way, or that they have been chosen, and that it is this special singling-out that enables them to perceive beyond the everyday. In fact, it is the aura that makes non-ordinary communication possible.

So while this type of communication is not everyday, neither is it anything special. It is a function that everyone possesses. The only difference between individuals is that some are aware of their aura's capabilities and some are not.

On the other hand, it is also true that many actively fear and suppress such communications.

The common unease with auric communications

When auric level communication is first initiated many people experience a great shock. They are confused by the signal. They may deny anything has happened at all. Then, if it is accepted that something out of the ordinary *has* happened, there is much uncertainty as to where the contact comes from. Those with a strong religious upbringing may be fear that such activity is not allowed, or even that the demonic is involved. For those with atheistic or materialistic conditioning, there may be no willingness to acknowledge that something they cannot explain has occurred. In each case, the contact is likely to be declined.

Psychologically, this unease can be said to result from unfamiliarity, given that unfamiliarity often leads to fear, denial and rejection. Such responses may be likened to the unease people feel when others around them are speaking in a foreign language. They wonder what is being said, and especially whether they are being talked about negatively. If their unease is acute, they may begin to feel threatened. Clearly, their discomfort arises out of their lack of knowledge of what is being said, and particularly from their ignorance regarding the intentions of those speaking in the unfamiliar tongue.

The same applies to auric communications. People reject, deny and fear them out of uncertainty regarding what is behind them, and especially because they are ignorant of the level on which such communication occurs. Fear is key. Fear of the stranger can be seen to be behind the discomfort people feel among those speaking in a foreign tongue. The same fear, bolstered by fear of the unknown, applies when people are faced with communications coming from non-embodied identities who use auric channels to disperse information.

Accordingly, the way to combat fear is to foster familiarity. That is what we will seek to do in this chapter, explaining why such communication occurs, and particularly how. With knowledge it is possible to become much more open to communication with one's own spiritual level identity.

Sources of communication

Human concern regarding the origin of auric communications is completely understandable, especially in the early stages, before evidence of goodwill has been established. With experience, and especially with affirmation that the communication is for the recipient's benefit, not detriment, trust in the source and confidence in the validity of what is being communicated can build.

Auric communications initiated by one's own spiritual self are the most common, the most frequent and, for many people, the most frequently declined. Simple uncertainty is the usual cause of that, allied to immersion in daily life that prevents the recipient from paying proper attention. Life activities can be so demanding that the recipient is too tired or unfocused to absorb the fact that communication has been initiated.

Most communications from the spiritual level to the everyday mind are to do with fulfilling life plans. As this type of communication has been thoroughly discussed elsewhere we will not dwell on it here.*

Another reasonably common type of communication emanates from spiritual level friends or guides. Such communications are often intended to provide a psychological jolt, what is colloquially called a wake-up call. This occurs when the individual is confused, has departed from the life plan, has descended into self-damaging behaviour, or has a significant life event on the horizon and the individual's own spiritual self is unable to bring what is intended to the individual's attention. There are many ways such a communication may occur: as a dream, as a shout in one's head, in the form of a vision, drawing attention to a book or image, a powerful feeling ... the list is extensive. What they have in common is that the communication is initiated out of goodwill, and the aura is affected in some way in order to draw the individual's attention to the act of communication. How is the aura affected?

The nature of auric sensations

It can be said there are two basic levels to auric communications. The first level involves attracting the attention of the embodied personality. The second involves the actual content of the communication.

An embodied individual's attention can be attracted in a number of ways. For the recipient, most initiations of contact are felt as a bodily sensation. Such

---

* For detailed discussions of life plans see *Practical Spirituality*.

sensations include a sudden feeling of tiredness, dizziness, heaviness, a surge of excitement or energy, an urge to play, sensations of heat or chill, the visual effect that the world is lighter or darker, a surge of sadness or happiness, or a feeling of agapéic ecstasy.

These are essentially calls to the recipient to pay attention. Once an established communication is in place, which results from repeated communications having occurred over an extended period, favoured means are usually then used to call the recipient's attention. This is entirely a matter of convenience, because the recipient comes to recognise a particular type of call to attention and so recognises who is calling.

Each of these supposedly physical sensations are not actually physical at all. Each occurs at the auric level. The calling spiritual identity presses the aura and the resultant sensation is tiredness, or a chill, or a surge of energy, and so on.

What is typically communicated?

Why do spiritual identities call on another who is in an embodied state? It is to help in some way. Often it is in response to a call for assistance.

Typical circumstances involving assistance include providing information, especially offering a different perspective, when a big decision is required. Another common circumstance is giving an individual who is feeling unhappy, depressed, lost, or uncertain an auric level hug, so to speak, so they feel they are supported and not alone. Often support is given to help an individual in relation to completing specific tasks related to the life plan.

Then there are those who communicate on behalf of others, who possess what are popularly called psychic abilities. They have actually established a means of consistently using their aura to communicate at an energetic and spiritual level with non-embodied identities who wish to communicate to human beings. This includes mediums who pass on messages between the living and the so-called dead (they are actually all living, just in different domains). Spiritual healers also fall into this category, as do those who channel texts such as this.

Because each person's aura has been associated with their body since their body was conceived, once a communication channel has been opened between the human personality and the spiritual identity information flows easily and naturally. It is just a matter of opening up one's mind to the possibility, stilling thoughts and emotions, and sustaining the intention.

It is not part of everyone's life plan to specifically focus on developing their ability to communicate on their auric level. Furthermore, and as we stated earlier, preparation is required. So when anyone engages in auric communications there has always been prior training that has prepared the individual for whatever is planned, whether that training has occurred during previous lives or has been carried out between lives.

So there is no point in forcing such matters. If you are attracted to engaging in auric communication, first enquire regarding where the interest comes from. Then, if you are satisfied it is a valid exercise, proceed in accordance with the guidance that will come to you. Guidance that you will need to ask for.

Is prayer effective?

Many people involved in traditional religions who seek advice regarding key life decisions are directed to pray for guidance. Prayer is insufficient. What is also required is to listen for a response. Responses are easily missed by those not practised at recognising when information is coming from outside their personality, especially when the response is quiet, as is usually the case, and the impulses of the body and personality are loud.

What is required is stillness. This creates a suitable environment in which the ordinary mind may consciously observe what responses arrive as a natural product of the questioning process. This applies whether prayer, meditation or any other process is used.

What is not widely understood is that a simple question and answer exchange may be established by which the everyday personality and the individual's own spiritual identity may engage in regular and quite straight forward communication. When this is established other non-embodied identities, particularly non-embodied friends and guides of various kinds, can add their input to the communications. At such a time a significant step has been taken in the node fragment's progress towards maturity.

What about unwanted "guests"?

A common fear is that when individuals open themselves up to auric level communication they will receive unwanted guests. Some go as far as to fear what are popularly called evil spirits will come calling. This fear is much exaggerated. There are certainly a number of categories of spiritual identities who might not be desirable. But they are not destructive.

One issue is that people become scared of ghosts and the like not because what they are in communication with is out to do them mischief, but entirely because the event is so unfamiliar that the perceiver responds with fear. This may be likened to hearing someone walking around outside the house and fearing they will attack you. Panic ensues, perhaps followed by a call to neighbours or the police. But when support arrives it is found the supposed attacker is just a serviceman doing their professional work. So the visitor's intention was benign, yet the recipient was scared witless. This often occurs when the disembodied, otherwise known as ghosts, are in the neighbourhood looking to communicate regarding their uncertain state. Much historical story-telling comes from such incidents, in which fear is conjured when no actual threat exists.

It is also the case that parasitic identities can associate with the human. This is like when a tapewom becomes lodged in the intestine. The tapeworm is not targeting the host for any other reason than for nutrition. It is uncomfortable for the host. And horror may result when the tapeworm is discovered. But there is no evil intent. It is just an animal seeking food in the way nature has designed it. The same applies to parasitical spiritual identities. And just as a pill can expel the tapeworm from the body, so directed intent can expel the parasitic identity. However, that identity is not evil. It is merely doing what it does.

Such parasites are never accidentally picked up. In the case of the tapeworm it is through handling food or material carrying eggs. In the case of parasitical identities it is through willingly and often ignorantly opening oneself up. Just as simple hygiene prevents infestation by tapeworms, so simple auric hygiene prevents similar infestation. We add that such occurrences are rare.

There is also much fear-mongering regarding people being possessed by evil spirits. In fact, most cases of possession or of multiple personalities are due to personalities lived during prior lives leaking into the current personality.

As far as malevolent identities at the spiritual level are concerned, they are as real as werewolves, vampires, and other much feared but entirely fantastical denizens of the night. That they are believed in so fervently by so many says more about humanity's ability to scare itself silly than it says about the reality of what is out there. What is out there is actually far more interesting than human imagination and fear allow.

We have discussed aspects of all these matters before and will do so again in detail in other contexts.

Establishing a demystified perspective

The majority of human beings are completely unaware that auric communications occur. That is useful in that it gives us a great opportunity to provide the wider populace, living in whatever location around the planet, with a new account shorn of supernatural elements. In harmony with our overall intention, we offer a simplified, demythologised, demystified framework to understand these phenomena.

We conclude by reiterating what was stated earlier in this discussion, that there is a class of sensation that manifests bodily as tiredness, or dizziness, or in a sense of not being quite mentally right. These sensations then suggest to recipients that they are physically ill or suffering mental aberration. In fact, these individuals are simply in receipt of a request from their own Dao-level self to pause a moment in their life and attend to the incoming impressions. Why? Because there is likely to be some benefit to them, and potentially considerable benefit.

To that we add the observation that what any person may conclude as a result of feeling such sensations is that they are in fact the embodied aspect of a greater team, all of whom who are equally alive and conscious, but are simply not in an embodied state, and that that team's intent is supportive and helpful.

Appreciating this is the case should enable everyone to feel, first, that they are not alone, second, that they are supported, and third, bouyed that assistance is available whenever they face the challenges that are inevitable during the course of a physically incarnated life. Finally, they can have confidence regarding the source of the ensuing conversation. For this is not at all about people and gods, but about individuals and their supporters.

We are all in this together

This web of individual support may be extrapolated into a much wider context, affirming the understanding that every member of every species is similarly supported, to the extent that it may be said, "We are all in this together." By this we mean that the web of life on this planet may legitimately be viewed as a cooperating whole.

Of course, there is a partial exemption to this, given there is competition as well as cooperation, to which may be added indifference, insofar as the planet is divided into environmental niches, some of which are incompatible with, unrelated to, and ignorant of, other forms of life. Nevertheless, the notion of an interlinked and cooperating web of life is applicable. The recently offered concepts of Gaia and the noosphere reflect what we are saying.

However, beyond individuals being supported from the spiritual level, we also wish to draw attention to the fact that there is a need, an urgent need, for humanity to embrace the notion that there exists a web of life. This means not just being supported by others, but for each human being to extend their support to those in the world around them in the form of environmental and ecological protection. It also means fostering every contribution and innovation related to ecological understanding. A drive to reduce the human population would offer pragmatic support of the notion that human beings are just one contributing part of the planet's web of life.

So this is a perspective that legitimises all forms of life, not just human life. It views all life as encompassed under one umbrella of existence.

The following chapters introduce the model of agapéic space, which offers a framework for thinking about the web of life and seeing how the human is part of an extensive range of inter-linked Dao-identities and bio-identities.

Agapéic Space and
Shamanic Flight

13

The Model of Agapéic Space

TO FIND ONE'S WAY across the surface of this planet it is helpful to have an orienting system. A basic tool was found for assisting travel when the lodestone was identified. This functions as a primitive compass, as a result of technology being developed that shaped the magnetic rock into a pointer. This was mounted on a fulcrum, which allowed the magnetic pointer to rotate. Subsequently, a coordinate system based on the magnetic axis was developed that proved useful almost everywhere on the planet.

The compass has had multiple applications. It was essential in establishing a map of the planet. It was also used to steer vehicles, particularly ships in the earliest times, and to guide travel over unfamiliar territory on land.

In early times the American Indians developed a coordinate system they called the four sacred directions. Although it is a coordinate system applicable to the physical sphere of the world, in its sacred sense it is also a means by which to orient oneself during shamanic flight. That coordinate system comprises four cardinal points and a knowledge that there is the middle world, which is this physical world, an upper world and a lower world, both of which can be travelled to during shamanic journeying. Each location is populated by a variety of identities. Accordingly, this is a system that utilises four cardinal points and three general realms of location: upper, middle and lower.

What we are offering here is a related and extended coordinate system we are identifying as agapéic space. It is related to ancient physical and spiritual mapping systems, but is somewhat more extensive in scope.

A definition of agapéic space

Agapéic space contains three sets of coordinates. These are agapéic frequency, willingness to bequest agapé, and hierarchy. This defines a volume comprising not just the human realm, but agapéic space for every species.

The purpose of these axes is to locate both species and individual identities within spiritual space, for this is the domain agapéic space maps. We note agapéic space is a *model* of spiritual space. It is a conceptual framework for explaining certain arcane aspects of Dao-identity as it exists in its natural non-embodied state within spiritual space. Just as a map is not identical to the terrain, so agapéic space is not identical to spiritual space. It is a description of certain aspects of spiritual space.

What is the significance of the word *agapé* in this description? Agapé is the nature intrinsic to Dao-consciousness. Loving regard, a desire to help, wishing the best for any other, self-concentration into goodness, willingness to love all others, and possessing an optimistic and positive outlook, are all part of agapé. Other attributes include a desire to nurture others, being willing to focus on any other individual so that other-interest predominates over self-interest, a capacity to hold a group in special relationship, willingness to see beyond the superficial personality to the core of another person's nature, recognising that any other is intrinsically and at core identical to oneself, and claiming commonality with all others, human and non-human.

A definition of hierarchy

Willingness to act from agapé produces hierarchy. Hierarchy constitutes the sum and product of loving acts enacted throughout a life. It is usually added up at the end of a life, which is the reason movement in hierarchy is most often assigned at the end of an incarnation—although some rare individuals may carry out loving acts of such power that their hierarchy is raised while still embodied.

For the sake of discussion, we designate hierarchy as comprising one hundred steps. Each step may be subdivided in order to accommodate an increase in hierarchy due to developing within each of a thousand lifetimes, but adding more numbers is not necessary at this point. We designate the one hundred steps as being a ramp or scale, literally a ladder of advancement. Jacob's Ladder and other similar metaphors reflect this notion of advancement in hierarchy.

Note that in defining hierarchy in these terms we explicitly disassociate it from human concepts of social hierarchy. Human social hierarchy confers status and social value, and as a result lifts one individual above another. Our notion of hierarchy contains no implications of status in any human sense. It does not place any individual node either above or below another node in any kind of human value-driven scale. Hierarchy is entirely a developmental term.

The axis we are calling hierarchy reflects an individual's incrementally increasing willingness to bequest agapé to others. Hierarchy increases as an individual experiences much, then uses both self-assessment and other-assessment to evaluate acts in each life, reaching an understanding of what was done in a loving manner and what was not. The individual consequently reformulates its inner nature to promote willingness to act from agapé in any and all situations, towards any and all living beings. It is entirely appropriate to imagine that a naive node fragment of Dao-consciousness has a hierarchy not of zero but of one.

The metaphor of hierarchy only approximately captures one aspect of what is actually a very rich developmental process each node fragment goes through during its experiences of human embodiment. Yet it clearly indicates that there is a consequence, a benefit, and so a reason to act in a loving manner.

The implications of agapéic location

Each embodied node fragment of Dao-consciousness has internal qualities that establish a natural location within the agapéic coordinate system. That is, each individual comprises a certain value of hierarchy, a certain value of agapéic frequency, and a certain value of intrinsic loving nature.

Given, for the purposes of discussion, that the hierarchy of each human being is essentially uniform for the duration of a life, the major feature distinguishing between individuals is their location on the two axes of agapéic frequency and willingness to bequest agapé. This results in the principal difference between human beings distributed across agapéic space, and explains why each feels more comfortable with some encountered embodied individuals and less comfortable with others. The following examples will make clear what we are referring to.

It may be said that one person feels comfortable with another because they are at a similar location in the agapé coordinate system. Equally, discomfort may result when with others who are at a significantly different location. Of course, other factors come into play, given people are attracted to others or repelled by them on many grounds, including bodily, emotional, psychological and professional. But in the case of long-term attraction, besides an obvious connection via a form of human relationship, an underlying reason that explains why people come to be at ease with one another is because they are similarly located in agapéic space.

Another instance is that members of the same spiritual cluster are naturally attracted to one another when they are in an embodied state. Certainly, this is in part because they are from the same family, to use that term, and so they tend to be naturally at ease with one another. But even family members squabble, as everyone well knows, and fall out. During cycles of embodiment karma may build up that leads to feelings of unease between those who are otherwise from the same spiritual cluster. However, it is also the case that those from the same cluster tend to develop the hierarchy component of their identities at more or less the same rate. So they are comfortable with one another on those grounds.

These doubled factors, of being from the same spiritual cluster and being at a similar position as measured by the agapé coordinate system, leads to individuals tending to come together socially and mixing harmoniously. Disharmonious mixing in a social group may occur because others who enter are at a different position in agapéic terms, or are from a different spiritual cluster.

This also applies to tight-knit spiritual groups. The tightness derives from them being from the same spiritual cluster. Others who join may be at the same or even greater agapéic development, but they feel excluded at a deep level and consequently leave. That deep level is at the spiritual cluster level.

Maturity and the agapéic scale

The purpose of incarnating is to use the experience to mature as a node fragment. The agapé coordinate system offers a way of measuring maturity within a clear and relatively simple developmental scale.

Generally speaking, an individual initially comes into a position within the agapéic scale corresponding to low hierarchy, low willingness to bequest agapé, and minimal agapéic frequency. The individual then progresses incrementally in terms of hierarchy, agapéic frequency and willingness to bequest agapé.

There is an end point on this scale. This is where incarnating identities become sufficiently mature that they no longer need to utilise embodiment to continue their development. We will provide further details on this presently.

To conclude, each individual node fragment matures in a way that may be described as a trajectory through agapéic space that is mapped by the agapé coordinate system. No individual trajectory can be predicted precisely, because anyone's trajectory is a consequence of choices made life by life. The point of graduation from the cycle of incarnations is, effectively, the sum of an individual's thousand or so lives.

When the node fragment graduates it is not the same node fragment. When it reaches the end point of the human phase of its trajectory the node fragment is not the same as it was at the beginning. Its graduation from the human zone within the agapéic scale is achieved in a literal as well as a metaphorical sense.

14

An Individual's Travel
through Agapéic Space

HAVING INTRODUCED THE NOTIONS of agapéic space and a co-ordinate system to map anyone's position within that space, in this chapter we will show how the model may be used by meditators as they negotiate spiritual space.

Two ideas are fundamental to any consideration of negotiating spiritual space. The first is that of having freedom to move. This may seem contradictory, given that in the previous chapter we identified individuals as occupying a fixed position in agapéic space, and said that individual's movement in agapéic space occurs only incrementally, life by life. But if a meditator is going to move through agapéic space, the implication is the individual is free to move. This tension between having a fixed location in agapéic terms, yet having freedom to move beyond one's fixed position, will be addressed in what follows.

The second fundamental idea is that there are other identities in spiritual space. So part of the whole idea of exploring spiritual space is not just to discover environmental niches, but to meet other nodes of Dao-consciousness of all kinds. Some of these nodes occupy a similar position to oneself in terms of agapéic scale. Others do not. How is it possible to meet others if you each occupy different fixed positions in the agapéic scale? We will also discuss this phenomena in what follows.

In ancient times meditators were known as shamans. The act of meditation, during which an embodied individual's awareness explored spiritual space, was called shamanic flight. In order to ground this discussion in human experience, and to show that what we are discussing is by no means new but has been practised throughout human history, we will use the terms shamanic space and shamanic flight. As we have repeated throughout this text, it is our terms and descriptions that are new. The experiences are not.

The source of this model

At this point, we need to clarify that this model has been developed in direct association with and by Peter Calvert. His capacity to visualise such matters has been used to structure the agapéic model. While he received little formal instruction in creating theoretical models during his education, nevertheless he was willing to cooperate with our intent.

Over a period of some years, during which he was driven by intense curiosity, we have been able to impart a workable theoretical model, which is presented out of our aim to stimulate some understanding of the hidden structure of spiritual space. Understanding one's location in spiritual space is important because it gives rise to confidence that, first, there is such a space in which to move, and second, that one has a place in that space independent of one's bodily location in physical space. This raises the issue of the relationship between spiritual space and physical space as it experienced through the senses.

There is no relationship between orientation in spiritual space and orientation in physical space. Orientation in physical space is conventionally described in relation to the axis of the physical planet. Obviously, spiritual space is not physical, so cannot be described in these terms.

Yet to say that there is no relationship at all between these two spaces is to negate their subjectively apparent relationship, which arises from the orientation of the body.

Correlating to agapéic space

The body has a aura. The aura has an associated hara-level structure comprising the dantians and the ida and pingala. An individual's orientation in agapéic space is coordinated using the hara-level dantian.

Thus the axis of agapéic frequency extends forwards from the hara dantian in the belly. The axis of willingness to bequest agapé extends leftwards and rightwards across the shoulders. And the axis of hierarchy is aligned with the spine. This correlation with the normal body structure and its orientation in physical space is the means by which a connection may be made between physical space and spiritual space as it is described in agapéic terms. [See FIGURE 4.]

In offering this model, we are asking that you use your imagination to project these axes onto the space that is experienced during meditation. However, we need to make clear that the experience itself is not imaginary. It is vividly experienced by practised meditators. It is also experienced by many during dreams, when their awareness is freed of its usual embodied limitations.

On the other hand, imagination can certainly impact on meditation experiences. Earlier we drew attention to the fact that people often have valid experiences, but then contextualise that experience using invalid notions, such as heaven and hell. The same applies to experiences in spiritual space. The actual experience can be overlaid by conditioned responses, as happens when a person sees an angel, a fairy, or Buddha, Krishna, or the Virgin Mary. They certainly perceive an identity, but conditioned thinking leads them to interpret their perception using imaginary terms. This response does not invalidate the experience. But it does invalidate their interpretation of what they experienced.

There is one other significant reason that individuals perceive differently when they explore agapéic space. This is due to natural and often unavoidable disturbances that occur in the correlation between an individual's bodily structure and shamanic space. This may happen because of tiredness, or because the body

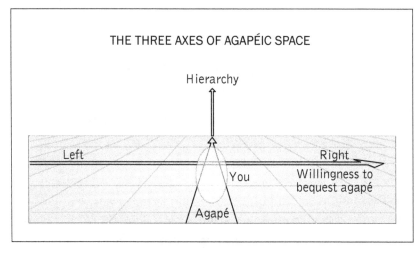

THE THREE AXES OF AGAPÉIC SPACE

Hierarchy

Left

Right

You

Willingness to
bequest agapé

Agapé

FIGURE 4

is reacting negatively to substances ingested to induce a experience of spiritual space in the first place. Whatever the cause, it leads to individuals losing their orientation in spiritual space, with the result that they do not know where they are in terms of what we are designating the three agapéic axes. This leads to an incorrect attribution of what they experience, which means that, in ancient shamanic terms, they designate an identity as situated in the upper world when it was actually situated in the lower world. In terms of the agapéic axes, they might wrongly designate an identity as being higher and to the right of them when that identity is actually on the same level but to the left.

Nevertheless, we maintain that the correlation between bodily structure and agapéic space holds, because this description is coming through an embodied individual and it is designed to speak to other embodied individuals. Because it is specifically linked to the human sense of orientation in space, the experiences of spiritual space had by spiritual identities co-associating with non-human species are not relevant to the description we are offering here. It is highly likely they would not find this model illuminating or useful.

Shamanic space and shamanic flight

Our model of agapéic space is an attempt to concretise a theoretical description sufficiently to satisfy a broad set of contemporary theoreticians and those who undergo this kind of experience. For our purpose, it may be said that agapéic space refers to spiritual space in its theoretical construct, while shamanic space refers to the explorer's subjective experience of spiritual space.

In using the term shamanic space you will note that we have chosen the most ancient, and therefore non-religious, terminology. It is not the realm of God, for God is a construct. It is not the realm of heaven, for heaven is a construct. It is not the realm of hell, for hell is a construct. Yet those ancient notions, which are mostly derived from experience, may be subsumed into the notion of shamanic space. We simply seek to make this an a-religious description.

It is difficult, if not impossible, to find ancient descriptions of shamanic space that agree well with one another. This is because so much historical material is lost, and much of what remains has been subjectively reconstructed long after the fact. We can do nothing about that. Nonetheless, it is into this confusion that we are attempting to introduce a little order so any investigator into what is conventionally termed spiritual experience may feel an increased level of confidence about what they are exploring.

We come now to the phenomenon of shamanic flight. As observed earlier, the shamanic world view commonly describes a lower world, a middle world and an upper world, and places different denizens in each. Yet traditional descriptions of shamanic flight do not clearly distinguish between the different modes of exploration. This is due to the relatively recent establishment of psychological language that is now used to describe personality and its experiences.

Descriptive language is so powerful that it shapes both an individual's self-identity and their perceptions. Each is interpreted through whichever language is possessed. A person in a different culture will use different language. The phenomenon of interpreting visualisations means that culturally-bound descriptions very frequently differ one from the other. This is due to the relationship between percept, interpretation and language.

The result is great difficulty in providing culture-free interpretive descriptions of spiritual experience. That is an ideal that will likely never be achieved. Nevertheless, adopting a single descriptive language enables experiences to be compared for similarity and difference, even if certainty that experiences are exactly the same will likely never be achieved.

With that clarified we now continue our exposition of shamanic flight within shamanic space. As a contribution towards establishing a common language to describe these experiences, we return to the three agapéic axes, to which we now add more detail.

The values of the axes in shamanic space

To remind you, the three axes that map agapéic space are agapéic frequency, willingness to bequest agapé, and hierarchy. They respectively extend forwards from the hara dantian, across the shoulder blades, and upwards from the spine.

The axis of agapéic frequency is subdivided into 65,000 levels.

The axis that denotes willingness to gift love, or as our collaborator has defined it, willingness to bequest agapé, is subdivided into 60,000 equal divisions.

The hierarchy axis is notionally divided into 100.

Taken together, the values of these three axes may be used to define a three dimensional space within which human exploration of shamanic space can be mapped. This volume is larger than any person can imagine. It is also compartmentalised into zones which are occupied by every species in all the universes. No species' zone of existence encroaches on any other species' zone.

Obtaining information by moving

Historically, there have been various descriptions of how spiritual level information may be accessed. These include the metaphor of the Akashic library, and the related notion that there are levels of access to the information it contains.

There is also the notion that an individual may develop various abilities by which to access information, such as telepathy, far-sightedness, or simple intuition, each of which facilitates access to information one desires or needs for oneself, or that one requests on behalf of others.

We suggest that information is obtained as a result of an individual being free to move. Previously we stated that freedom results when one acquires hierarchy, and hierarchy is acquired through being willing to bequest agapé. As a consequence, one migrates sequentially, by progression, through putative realms of existence, each inhabited, as we have just stated, by various populations.

In the earliest phase of the novice node of Dao-consciousness, when it has not yet experienced physicality, there is not much freedom to move through levels of spiritual space. As experience is acquired, that node (or node fragment) acquires increased freedom to move through the levels, although only temporarily. An assignment or quality confines individual spiritual identities to a particularly located realm until they transcend it.

One could conceive of these as contradictory attributes, in the sense that there is a fixed location, but a degree of freedom. Yet as the identity progresses through layer after layer of realms, freedom increases.

Freedom is earned. In that sense, it is directly related to hierarchy. Also as with hierarchy, freedom has a developmental aspect, related to the acquisition of knowledge regarding aspects of existence in each domain. This knowledge is acquired from experiences undergone during successive incarnations. As the information is processed and understood, it promotes greater freedom.

As a result, a node fragment (to speak of human incarnation) incrementally advances through the levels mapped by shamanic space to a condition of much greater freedom of movement. This is due to its knowledge increasing geometrically. Note that the relationship of knowledge to experience contributes to a scale that is geometric, not linear.

The development of a spiritual identity occurs as a direct product of the information it acquires through incarnation, information of whatever type, wherever derived, and via co-association with whatever species. That information is subsequently uploaded to the individual Dao-identity. Each such accumulation of information, after it has been duly processed, enhances the degree of freedom possessed by the individual. We observe there is an additional energetic component to this process which cannot be described within the model as currently constructed. So this description is necessarily incomplete.

Overall, the parameters we are discussing are only discernible at a subtle level. Yet this is sufficient to identify some general tendencies in relation to shamanic space, particularly regarding its variable character and the variety of occupants one may encounter within this coordinate system.

We offer this information as a hint towards a deeper understanding of the nature and attributes of a node of Dao-consciousness.

Freedom as values on the agapéic axes

When a node fragment of Dao-consciousness reaches a sufficient stage of maturity it no longer needs to incarnate. We now identify the growth to maturity of the node fragment in terms of the three axes of agapéic space.

Accordingly, maturity on the scale in hierarchy means achieving 100%.

Maturity on the agapéic frequency scale involves progressing from a frequency of 25,000 to a frequency of 35,000. These numbers denote the range humanity occupies on the agapéic frequency scale. We will discuss this shortly.

Maturity in hierarchy involves sustaining a movement involving an undisclosed number of acts of willingness to bequest agapé. We define that movement as involving progress from 41,000 to 45,000 in the scale of willingness to bequest agapé.

There is no particular reason to assign these numbers. They are convenient. These axes are a construct, after all. Positioning the human in the centre of the model is merely expedient. So it is essential that readers appreciate the model is representative rather than factual, being offered as a frame of reference of use when discussing experiences. Agreements obtained by sharing descriptions of experiences with others who have participated in the same experiences is of fundamental importance in building mutual confidence that yes, similar perceptions were received, yes, emotional and energetic perceptions were similar, and yes, they have been interpreted in similar ways. It is then possible to confirm one another's experiences. The more people are involved in such mutual affirmation, the greater confidence can be derived.

Yet even that is only partially true. This is a descriptive tool useful for ascribing relativity, one human being to another. It has little use beyond that activity.

15

The Range of Beings
on the Agapéic Scale

HAVING EARLIER EXPLAINED the axes of hierarchy and willingness to bequest agapé, we now turn to the third axis, that of agapéic frequency. Agapéic frequency has two primary uses.

First, it denotes an individual's developmental progress towards maturity within the human domain. We assigned that to a movement from 25,000 to 35,000 on the notional agapéic frequency scale. The second is that it situates the node fragments that co-associate with humanity into a much wider conceptual framework. Because we discussed the first aspect, that of mapping personal development, in the previous chapter, in this chapter we will focus on this second use.

Expanding the concept of agapéic frequency

A primary intention behind the agapéic frequency axis is to indicate the natures of nodes of Dao-consciousness as they co-associate with all biological species on this planet, not just the human species.

The nodes that are cast out of the Dao possess different levels of complexity. When they subsequently seek to co-associate with a physical species, nodes select a physical species that has a level of bio-complexity that matches and is appropriate to their Dao-complexity.

In this chapter we will not discuss the nature of non-human co-association in any detail. Our purpose here is rather to indicate in broad strokes the basic differences between nodes co-associating with the physical domain on this planet using the scale of agapéic frequency.

We previously introduced agapéic frequency as having a notional range of 1 to 65,000. For shorthand purposes we will refer to this range as being from 1 to 65. Having clarified this, we will outline the frequency range of nodes in very sweeping terms, beginning with vegetation.

Overall ranges on the agapéic scale

The range of agapéic frequency appropriate to the range of species identified by the English term plant, we nominate as being between 5 and 10 on the agapéic frequency scale. Note that this scale does not apply to plants, but to the nodes of Dao-consciousness that co-associate with plants. As this is a broad outline only, we will not describe in any detail how nodes actually co-associate with plants.

Group or hive minds, for instance of vertebrates such as aquatic species, avian species, and nests of social insects such as ants, co-associate with nodes of Dao-consciousness that exist in the range of between 10 and 15 on the agapéic frequency scale. We also place invertebrates such as worms, and those with exo-skeletons, such as crustacea, within this range of nodes.

Those nodes of Dao-consciousness that co-associate with warm-blooded mammals we place on a scale of 15 to 20.

Elementals, those non-embodied nodes that oversee vegetation and environmental niches, we situate from 20 to 25. We will say more about them shortly.

The node fragments that co-associate with human bodies extend across a range from 25 to 35. By this we mean that when an inexperienced node fragment first associates with a human body it is at 25 on the agapéic frequency scale. As it matures its agapéic frequency increases. When it reaches 35 on this scale, and when it also achieves maturity on the other two axes, it graduates from the incarnation cycle and rejoins those other fragments at the same developmental stage who are beginning the process of reintegrating the fragments into a reunified whole.

The cetacean family of whales, dolphins and porpoises co-associate with nodes of Dao-consciousness that extend from 20 to 30. So there is an overlap with humanity that makes it possible for node fragments of this complexity to migrate to human experience, although not many choose to do so due to the demands the human body and human interactions place on them. It is actually more common for node fragments co-associating with the human to spend time among the cetaceans. Although this too is infrequent. Most of those rare individuals who seek an alternative body form and social experience select species from other planets.

Note that, like the human species, there is a frequency range for the nodes that associate with various forms of physical life. As with the human species, this range indicates a developmental scale. That is, nodes begin co-association at the lowest point in the range of agapéic frequency, and as they mature they progress to the highest point in that range. They then graduate from embodiment.

We situate ourselves, as a reintegrated node of Dao-consciousness, at 42 on this scale. Given that we were once embodied node fragments, and that we have previously progressed from 25 to 35, what this model implies is that nodes of Dao-consciousness are not limited in their evolution. As they experience and learn, so they themselves expand their capacity, in most cases to a radical extent, particularly compared to their naive and inexperienced state when first cast from the Dao. Hence this scale applies only to nodes during their cycle of co-associating with a physical species. They all grow beyond that. The range beyond 42 we will not discuss as this is not our brief in relation to this transmission.

The other significant aspect of this scale is that it relates spiritual identities to physical species and shows that the human is not the sole conscious being on the

RANGE OF OCCUPANCY ON THE AGAPÉ SCALE

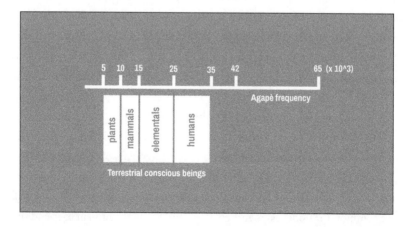

FIGURE 5

planet. The human species is not the sole repository of what is traditionally called a soul, which we are naming Dao-identity.

Every living creature has an associated spiritual identity. *Every* living creature. So when human beings indiscriminately cut swathes through species on this planet, they are not just being murderous in relation to other living creatures, they are also being disrespectful to other nodes of Dao-identity who, like human beings, are using embodiment as an opportunity to learn and evolve.

In addition, we have characterised humanity as extending ten points on the agapéic frequency scale, and all other species as ranging only five points. What this means in practice is that humanity's agapé qualities, especially its willingness to bequest agapé, is more deeply buried than in other species. Those who have pets will readily see the truth in this statement. So it means humanity has to work much harder to learn agapé. Giving all other species the freedom to occupy their niches without human impingement, over-exploitation, or worse, provides an appropriate opportunity to start expressing agapé. We will return to this in the final chapters.

With that stated, we conclude our exposition of agapéic frequency. To repeat, this scale is designed to create a numeric mapping of co-associated nodes and species. It is a somewhat crude and even arbitrary device generated for instructive purposes. But it means that the three axes of agapéic space can be applied to all species on this planet, not just to humanity, and their relative positions mapped. We will now comment on elementals.

The nature and purpose of elementals

In some historical literature spiritual identities are referred to as nature spirits. We choose to call them elementals. These aspects of identity are simple and small in nature, but they contain the Dao and express agapé as part of their intrinsic nature.

Some elementals care for and nurture plants and trees. Their care does not extend to insects, for they have their own associated Dao-identities. Plants have sentience, but no more. So elementals—who have traditionally been called gnome, fairy, elf and suchlike—act in love, being dedicated, observant and patient as they nurture and attend to members of the plant kingdom.

Time is not a function of elementals' awareness, for time pertains to physicality not to spirituality. Elementals are as constant in their attention to those aspects of the created world they have in their charge as any other spiritual identity functioning at any level.

One could view the elemental as a fine structure or local manifestation in the Dao, with an existence that begins and ends. In that sense, elementals have a lifetime. They come into existence and die, which is the end of their manifestation. And yet, because they are a feature of spiritual life in the same sense that humanity is a feature of spiritual life, it is appropriate to acknowledge their deaths alongside human deaths, for they are no less significant.

Human mythology regarding elementals

We would now briefly remark on humanity's mythologising of elementals. This category of being has been observed throughout human history. Humans have consequently projected onto them aspects of their own nature, particularly psychological characteristics such as tendencies towards greed, hatred, avarice, jealousy and similar.

The fact is, these beings have never possessed the characteristics humanity has projected onto them. They have love at their core, which is expressed in their purpose, which is to nurture those plants with which they co-associate. They hold no animosity for any order of created being.

As their purpose is not directed towards humanity, they have no interest in humanity, even though perceptive humans have sometimes observed them and told stories about them. But, as just noted, the stories reflect human obsessions, and do not reflect these identities' actual nature.

Elementals exist in varying magnitude. Any discussion of magnitude requires a reference point. In the context of this discussion we offer as a reference point the dimensions of a human being—not in their physical form, but their auric form. Accordingly, we take a two metre value as representative. This is naturally approximate and generalised, but it provides a way to consider the dimensions attributed to spiritual identities. Accordingly, taking two metres as a unit of 1, elementals typically range in size from 0.1 to 200 such units. Perceptions of identities outside this range do not involve perceptions of elementals.

We choose this range of perception purely for convenience, as a conceptual tool. It has no other application or purpose.

Contacting beings on other levels

Under special conditions, human beings can and do perceive elementals. This is because they are adjacent to humanity in terms of agapéic frequency, and so able to be perceived under appropriate conditions.

When an individual has the capacity to move with freedom up and down the agapéic scale, such identities can be perceived, met and interacted with. Their characteristics may then be noted in terms of energy signature. There are examples, available in accounts of Wisdom School group meditations, of interactions between meditators and nodes associated with canines, in which mutual awareness occurred. However, such incidents are rare.

It is the mobility of those meditators that enables them to make contact and associate with identities on other levels of the agapéic scale. As a result of their movement up and down the agapéic scale they encounter different orders of beings who, through their own capacity for movement, have elected to come to the level occupied by those meditators, or to come part way towards them, and thereby attract their attention. This association on a similar level makes communication possible, temporary though it may be.

In the same way, this mobility has enabled some meditators to experience rather different orders of being that normally occupy ranges other than the human agapéic spectrum of 25 to 35. By that means they encounter species such as canines and elementals. It also includes orders of being whose task it is to interact with them in order to convey this transmission.

Updating the Great
Chain of Being

16

The Great Chain of Being

CONTINUING THIS PROCESS OF clarifying the relationship between the spiritual and the physical, we now propose to examine an artefact of historical literature. We will confine our comments to the sixteenth century Christian and Catholic understanding as it is summarised within the model of existence called the Great Chain of Being.

The influence of this model continues to the present day. In fact, given that the postulated elevation of spiritual influence establishes a polarity between higher and lower, the model is applicable to almost every religion. This influence is associated with and predetermined by the shamanic experience of flying through the upper world and meeting its inhabitants, which act is itself the result of meditation experiences undergone throughout history.

We have in mind a formal treatment, in which we will distinguish between the hierarchical model proposed by the Great Chain of Being—which situates God at the top and devils at the bottom—and our proposal for a modern modification, in which ascension within the Great Chain of Being is conceived as being lateral as much as, or even more than, vertical.

We begin by differentiating between spiritual movement and human social hierarchy. Dominance, as a human-centred concept, is associated with height. In contrast, spiritual progression is a function of increasing experience throughout lifetimes, as is depicted in the model of progressing through agapéic space.

Spiritual movement and social hierarchy are fundamentally distinct phenomena. A by-product of acknowledging this distinction is to attack the myth of god-kings, and to detach the potential of achieving freedom from the concept of divine right. The notions of god-kings and divine right are purely human inventions that are not applicable to the spiritual realm.

Fray Diego de Valadés' version of the Great Chain of Being, published 1579.

FIGURE 6

A background to Valadés' Great Chain image

The version of the Great Chain of Being we intend to examine was created by Fray Diego de Valadés and included in his book *Rhetorica Christiana*, published in 1579. The book was completed after much work and was hailed as a masterpiece of exposition.

The word *rhetoric* in the title was understood in those days to indicate that this was idealist and promotional literature. The book was designed explicitly as a teaching tool to convey meaning to the masses, yet also be sufficiently theologically robust to withstand critical examination. As a product it was widely valued and utilised, providing materials for sermonising to the intended Mexican population, and to others. It was generally agreed to be a successful tool, providing a diligently constructed exposition.

Today it serves a different purpose, as it provides a window into another time and culture, particularly Christianity within the Spanish culture as the Catholic Church went about its work of indoctrinating Mexicans. So *Rhetorica Christiana* has been in every sense an influential document. That it is no longer so is a simple consequence of cultural change occurring over centuries.

Propagandising a religious world view is as much about the human desire to generate social hierarchy as it is about sharing a specific way of viewing the world. The so-called intelligentsia claim, and are given, high status in every culture. With political support, they make great public shows of piety. Their purpose is to impress and enlist the support of the uninformed, who are duly impressed by ritual displays. Thus a population is conducted into compliance through example. The fact that the example is staged, choreographed and manipulated by sophisticated marketing tools is more apparent to the lower when they have less fear of the socially and religiously higher. Hence fear is a very common tactic used to support and magnify the rhetoric. When—as is usually added to the mix—certain people in the populace are identified as lesser beings, then actions that generate fear can be used without the perpetrators feeling their consciences disturbed. By these means a naive Mexican population was subjugated by an invading force, itself supported by weapons and armour, and driven by the intention to evangelise.

This unholy alliance between dominant belief systems and the desire to dominate politically creates business opportunities within an expanded empire, a goal that has been the basis of every sojourn into foreign territory by colonial powers, no matter when or where. Domination is ensured by the displacement of cultural leaders, achieved via force if necessary. This tactic is sufficient to overpower the population. Dominated, they are put into a defensive condition. The physical postures of the dominated often convey their situation in body language that is invariably understood across all cultures. However, these subtle expressions are generally lost on the invaders.

The reason Valadés produced his imagery

Valadés was a Mexican monk. He created his Great Chain of Being imagery because he considered his fellow Mexicans needed pictures to more easily understand the newly proselytised Christian religion. Valadés was convinced his countrymen should dispense with the traditional understanding of Mexican culture and adopt the "good news" of Christianity. Accordingly, he summarised his understanding into the sections of his image. In this sense, he added to the wider Christian promotional literature.

The Great Chain of Being is a metaphor that underpins most current religious thought. It represents acquired understanding constructed from spiritual experience, often gained in meditative states. However, it also includes much that is historically constructed. So the ideas it contains are a mix of experiential knowledge, guesses, and extrapolations from sensory experience. Further ideas subsequently accumulated around those extrapolations, which, from a twenty-first century perspective, are obviously false.

The concept of the Great Chain of Being was archaic even in Valadés' time. It represented understanding that had been accumulated over hundreds of years. Nevertheless, it is representative of the best of intentions, even if the social purpose of this intention was to share this supposedly love-based set of metaphors in order to gain control of Mexican minds.

The image attempts to persuade that it is comprehensive by referring to all the then known and imagined orders of existence. It focusses on living creatures and places them in a hierarchy. That hierarchy runs from the least animated, possessing the least obvious independent freedom of movement, and proceeds through the various classes of organisms occupying the primary environmental niches of land, sea and atmosphere. It places humanity at the top of that section of the Great Chain. Angel imagery, stimulated by records of humanity's subtle perceptions, is placed above the human. The notion of spiritual hierarchy is encapsulated in the upper section, where it progresses through archangels to beliefs about the highest glory encapsulated in the idea of Godhead.

Unfortunately, there are many problems with this ordering.

The problems with Valadés' imagery

First, Valadés' illustration is predicated on human social hierarchies. Second, it makes the assumption that animal species are ordered in significance, with some, including the human animal, being given more privileged positions. Third, it privileges the spectrum of spiritual existence as being of higher significance and greater privilege than the entirety of embodied species. Fourth, there is nothing to indicate that what is being addressed is a spiritual hierarchy rather than a mixed spiritual and embodied hierarchy. Fifth, there is no indication, apart from a very subtle series of lines, to imply that there is direct influence from the highest level to the lowest. Sixth, the entire document is a carrot and stick model, indicating to the human that some things are best aspired to and others best avoided.

Valadés' exposition of the Great Chain of Being also involved an attempt to describe the Ground of All Being. Using our own terminology, we identify this Ground with the Dao, the ultimate unmanifest. However, Valadés' attempt to connect the Chain and the Ground was unsuccessful. This is because his imagery does not properly differentiate between the Ground and the figure.

The various sections, subdivided within the overall image, each represent a figurative aspect that is pictured as distinct from the wider context in which it exists, that being the Ground of All Being. Thus the illustrated sections show species of many kinds that have developed out of the Ground. But they are illustrated with no distinction made between their physical organic bodies and their indwelling spirit, to use that language. The imagery provides no acknowledgement of the consciousness developed within them.

It is in contrast to this that we have postulated an enhanced model incorporating a description of a purely spiritual hierarchy, comprising a description of shamanic space suitable for the twenty-first century. The description of movement in ascension is better facilitated by a three-dimensional frame of reference, establishing the idea of movement through an imagined space, which is itself defined by vectors. Within that description the specified parameters provide a theoretically unlimited conception which is not bound to the example of this particular planet, but is applicable across all universes. It also provides a sound relationship to the Ground of All Being, the ultimate unmanifest.

We will now consider the segments of the Great Chain of Being.

Segmenting the Great Chain imagery

The spiritual hierarchy is illustrated clearly in the section above the human realm. There are angels apparently singing in adoration. There are small flying baby-like creatures representing the host engaged in child-like love. There are archangels whose only task appears to be to focus on and bring perfumed clouds of fragrance into proximity with the highest level. The principle of spiritual support is represented by the chain itself held in the right hand of the God figure, whose strength is implied by the mass of individuals all being supported by this one chain. The Jesus figure, apparently in recovery from the trials endured during his life that ended on the cross, rests in the embrace of the God figure. And what is presumably mother Mary is also at that level, hands in supplication, presumably for her son.

The porcupine-like spikes of energy, flinging droplets of radiance all around, serve to emphasise the high energy condition of that level within the clouds. The implied brilliance is a metaphor for the supernal light commonly experienced across all religions as emanating from that level. Timelessness is indicated by the lack of death, with nothing other than continuous glory depicted at this level.

Underneath the highest sections, and progressively figured down the chain, are humans, birds, fish and crustacea, and animals, both real and mythical. These represent diverse species in the primary modes of air, sea and land, collectiavely all the beings who occupy ecological niches.

The separate region of plants at the bottom conveys the understanding that the hierarchy of worth implied in this image is also a representation of the ecosystem as it was known in those days. Food chains very often contain the plant world, as direct sources of photosynthesised energy are the foundation for life-forms at higher levels.

The exception to that is the bottom zone. Ignoring the fantasy of subhuman cruelty, the prime representation is of flames, implying strong heat. We contend that the ascetic practice of meditation, and the experience of feeling heat from beneath, as it were, is the entire basis for this representation. Artistic licence is responsible for the linking of highly uncomfortable heat to nefarious practices and illegal and immoral activities, using horror to provoke fear, even terror, in the mind of the viewer.

The fact that the chain descends to that level implies that the practice of evil is a product of the all-loving godhead, which is plainly ridiculous. Nevertheless, this powerful composite image conveys the story developed from the time of the emperor Constantine, drawing on a circular logic and generating a distorted view of reality. The circular logic runs as follows.

The directive is that if you are good and exercise love you will not be hurt. But if you fail to be good or to exercise love, in the worst instance you will fall into the invented realms of hell where you will undergo the worst kinds of torture. The error is that the proposed godhead actively supports that worst outcome. So the choice remains with the individual. If individuals are well-informed about their choice, the implied expectation is that no one would ever actively choose that worst result. Nevertheless it is self-evident that some do, for there is a population in that worst place.

In contrast, the understanding we wish to convey in this fresh start is that everyone naturally has choice. Where an intention exists to investigate worst outcome, as every person does, there are more or less predictable consequences. But those consequences comprise nothing resembling the heat realm within the Great Chain of Being. Yet in one sense it can be perceived that while the illustration is indicative and inaccurate, neither is it wholly wrong.

17

The Great Chain and Agapéic Scale

HAVING ESTABLISHED THE NATURE of Fray Diego de Valadés' composite image of the Great Chain of Being, and having satisfactorily segmented it into partitions, we will now comment on each partition.

Angels, cherubs, animals

The European conception of angels as non-human is a misconception. In fact, angels are human, and are best regarded as aspects of non-physical human identity. This accounts for angels' interest in the human, because generally what is religiously considered to be the angel assigned to each human is in fact the fragment of a node of Dao-consciousness, comprising the higher self where prior personalities are collected. Therefore it is to be expected that an angel's prime interest is in the individual embodied human, specifically, its identity, its formation, its survival for the duration of its intended sojourn into physicality, and its challenges and contentment while shrouded by the physical form.

The aspect termed cherub may be interpreted as fragments of other nodes of Dao-consciousness, which supervise orders of life. We framed our comment in this manner in order to disrupt human certainty and established ideas concerning such things. This is because a node of Dao-consciousness embodied within a particular species is not necessarily what is expected.

In asserting this, we point to the range of agapéic frequency notionally assigned to individual members of species other than the human. In general, the nodes of Dao-consciousness that coalesce with most species do not fragment before doing so. Their character is a simple order of loving nature, and they undertake their task with a willingness and constancy that is a natural attribute of their nature.

In relation to animal species, a node of Dao-consciousness will take on a supervisory role in relation to individual animals. This occurs in the same way that an individual human being is "supervised" by its Dao-identity. As a result, a number of individual animals are overseen by an appropriate node of Dao-consciousness. This occurs in the exact same way that a succession of personalities embodied in individual human beings are overseen by a node fragment of Dao-consciousness.

In each case the organism acquires experience by virtue of living within its environmental niche. That experience is subsequently recorded into the associated spiritual identity. In the case of simple organisms, experience is accumulated across multiple individuals and then uploaded at the death of group members. This is paralleled in the human, for whom accumulated experiential data is uploaded when an individual dies.

With simple organisms experiential data is averaged across multiple individuals before being uploaded. This is merely convenient. Yet this process also reflects the density of information, to phrase it that way, which is accumulated by simpler organisms living smaller lives—that is, when compared to the more complicated lives experienced by more complex organisms such as the human, horse or cetacean.

Plants

What we have just said pertains to all varieties of animals, including birds, fish and crustacea, but does not apply to plants. Plants are a special case in the sense that, due to their lack of mobility, their experience is limited to encountering those members of similar or different species in close proximity, within reach of their branches, stems or roots.

Their awareness is similarly limited, apart from their capacity, on the molecular level, to sense and recognise what is contributed to them by their local environment, or, if the contribution is wind or water-borne, from a greater distance. This means a variety of responses are available to them, according to the molecular or ionic signals they receive from the environment via absorption.

Many species have evolved effective strategies to compete with each other. Fungi, for example, are in constant competition for the chemical resources of their environment. Ant species constantly interact with chemical trails and wind-borne scents, as do many other insects.

Fish and crustacea are endowed with both signalling and sensing techniques, mostly chemical and acoustic. All such organisms acquire information from their environment, generating a history of their reactions to their locality, and either surviving, thriving or expiring accordingly. This information relates to the quality of match between the organism and its niche. This is meaningful information that is collected and collated on the non-physical level.

The information is collected, and each plant species is supervised by an appropriate and willing node of Dao-consciousness. We earlier identified these nodes with elementals. Their purpose is to optimise organisms' formation and continued existence in the long term.

The agapéic scale extends the Great Chain

The model of agapéic space can now be viewed as an extension of the Great Chain of Being. As we have noted, while the Great Chain of Being has some strengths, it is overly simplistic. Not being able to map the human on the spiritual level, or picture its development spiritually, is a major failing. Another issue is that it lacks any indication of the role and goal of ascension within the human domain.

We have developed the model of agapéic space in order to address this failing. We have also designed the agapéic model to account for all species, not just the human, in order to illustrate the relationship between different classes of species, even allowing theoretically for the inclusion of populations of species that exist outside this particular planet. And we have encoded more detail in the agapéic model concerning the human realm than has previously been available.

The term *realm* itself is used here in both the micro and macro senses. On the macro level, the major steps within the agapéic model provide for the allocation of entire species, including animals and plants. On the micro level, opportunity is provided for mapping changes in location of individual identities. These changes in location reflect the different stages of their ascension.

Our use of the word ascension now requires explanation.

A new interpretation of ascension

The term ascension is loaded with historical baggage. We seek to separate this new description from that baggage by introducing fresh terminology. Accordingly, and as a new definition of ascension, we offer the phrase *incremental migration in agapéic space.*

We have phrased our definition this way to avoid it being co-opted into those errors traditionally associated with ascension—in particular, to avoid the possibility of any organisation seizing on it as a self-promotional opportunity and subsequently claiming credit for rapid advancement, by leaps and bounds as it were, through the ten thousand levels, and thereby missing the detail. We have already spoken at length concerning the criteria by which advancement through the levels is determined and need not repeat them here.

Describing advancement by referencing numbers on the relevant triple axes of agapéic space is a matter of simple numerics. It has its corollary in movement in three dimensional space.

The divine is a false construct

That completes our expansion of the one-dimensional depiction of spiritual hierarchy offered by the Great Chain of Being into the three dimensional model of agapéic space. Given that this entire description is centred on and oriented towards the human, more details involving other species are not necessary.

Instead, an additional detail will be introduced at this point. This is that the entire human concept of the divine is a false construct. We will begin clarifying what we mean by distinguishing between height and hierarchy.

The following is an observation commonly drawn from human experience: "When I begin my life I am small. But as I mature I grow, until at last I reach the height of the full adult." Even if that fully grown adult is short in relation to other adults, it is still accorded adult status.

As a result of referencing physical growth in terms of height, human beings have a natural predisposition to use height in other contexts, particularly when considering the social status of one human relative to another. So one person is described as possessing higher status, the other lower status. But this human predisposition may naturally lead to misunderstanding regarding our use of the term height in relation to the agapéic axis of hierarchy.

A typical example of misunderstanding occurs when a meditator subjectively perceives an incoming identity as being at a greater elevation. Human beings are predisposed to merge height and social status. The meditator then applyies this assumption to their perception, concluding that the incoming identity must have greater hierarchy, in a spiritual sense, than the meditator. The perception of greater elevation leads automatically, entirely as a result of imaginative projection, to the conclusion that the incoming identity possesses greater spiritual status. This connection is not valid.

Accordingly, we wish to clearly distinguish between a perception of height on the one hand, and height as we have metaphorically applied it when devising the term hierarchy. Maintaining this distinction is important for anyone who wishes to use this theory to understand their own experiences.

Any embodied individuals who are capable of such perceptions will, from their perspective, perceive a high level spiritual visitor either as coming from a higher level, or of projecting their presence downwards in order to interact with the human perceiver. This is a perfectly valid and true perception. But it is not true that that perception therefore defines the value or status of the incoming visitor. It is not true that the visiting being is intrinsically different from the perceiving meditator.

Historically recorded accounts of such perceptions are responsible for the myth that underpins the Great Chain of Being, leading to the mistaken view that there is only one axis and that it is vertical. Our description, utilising vector relations within an axial framework, provides a somewhat more sophisticated understanding of a spatial volume, in which any identity can be positioned and described in terms of measurements in relation to the three axes.

Simply, we offer a rational descriptive tool comprised of multiple dimensions in order to promote a more sophisticated understanding of human spiritual level perceptions, as they are experienced in the phenomenon of interpersonal relations that take place in shamanic space.

Conceptual outcomes of the new model

Accordingly, we assert that the path of ascension is not just vertical, but may be described in relation to the three orthogonal axes. There are several reasons for doing so, and a number of related outcomes we seek to make clear in relation to them.

The first outcome, as we have just stated, is to disengage human understanding from the simplistic one-dimensional model of hierarchical ascension.

The second is to take the image of Godhead, which is traditionally placed in an ascendant position in relation to humanity, and displace it beyond human perception in a horizontal rather than a vertical direction.

The third is to disconnect ascension from any ideas around social hierarchy, especially as fuelled by every child's experience, and to consolidate our claim that spiritual reality is fundamentally distinct from the simplistic notions contained in the model of the Great Chain of Being.

Fourth, to state emphatically that all embodied persons, whatever their social status, are fundamentally equal in spiritual status.

Fifth, to deflate and debase concepts of spiritual hierarchy that have been historically structured into religious hierarchies, whatever the religion may be.

In promoting these outcomes, we seek to decontaminate current spiritual concepts and show that ascension has nothing to do with human social structures that foster notions of hierarchy and status. Any association of ascension with status can only be viewed as the spiritual being contaminated by the social.

This is not to claim that the spiritual domain is special or different or pristine, or anything else that is traditionally ascribed to the so-called divine realm. Rather, it is to clarify that the spiritual is simply another location humanity can occupy, of no greater intrinsic validity or worth than the physical domain and its associated human social life. Hence, a prime reason we offer this refreshed description is to deconstruct traditional notions of divinity and elevation.

Historically, human understanding of the character of the non-physical realm occupied by humanity has been blurred by conflation upon conflation, to the degree that it is unrecognisable from the perspective of the individual when out of the body. It is laughable that human beings should dupe themselves into making such distinctions between what are simply alternative domains in which experience may be sought.

Individuals seeking power, prestige, status and wealth have mutually contrived to raise themselves above the ordinary human population and to maximise their advantage over those they would lead. This should not be surprising. It is a natural human characteristic to seek competitive advantage.

Throughout history, stories have been manipulated, both in relation to how the invisible realm was conceived, and in the ways public homage was offered to that realm. The unlettered, uneducated individuals who made up the ordinary populace historically were impressed by the apparent humility of those who led the homage. What they failed to see was the greed that drove their leaders. We deliberately do not hold back from using plain language.

The evidence for our perspective can be seen in the wealth accumulated by the religious who make such claims, and in the ways they seek to increase their followings and so further increase their wealth. The sophisticated arguments formulated throughout history, which have been constructed to multiply these advantages, are further testament to the degree to which the characteristic of being disembodied has been given a grossly inflated description for venal purpose.

Ascension redefined

Given this is the historical context, it now becomes clear why an alternative view is required. Accordingly, it is possible to say, yes, there is an identity that is non-physical in nature and which may be described in various ways. It makes a lateral transition to accommodate itself within a chosen species. When its experience has accumulated sufficient information, or at the demise of the member of the species concerned, it again moves laterally, at least in conceptual terms, and processes the experiences it accumulated while embodied. As a result of doing so repeatedly, it accumulates more and more information. So it becomes more knowledgeable. But throughout these repeated incarnations, its aim is not to increase its status in the ways that so occupy humanity.

This is not just an issue for human beings. It occurs within all hierarchical social species. In every case, periodic corrections are required to decontaminate the population concerned. By this we mean there is a need to remove accumulated misconceptions that skew perceptions, cloud understanding and confuse descriptions.

Given all this, we restate our intention to decontaminate traditional descriptions of spirituality, first, by seeking to return such descriptions to the reference point of shamanic experience; second, to associate shamanic space with the model of agapéic space; and third, to detach this fresh description from traditional Axial Age religious descriptions.

18

Sustaining the Web of Life

WHAT IS USUALLY FORGOTTEN in debates about religion, spirituality, science and environment is that all beliefs originate in firsthand experience. We previously noted how spiritual perceptions contributed to the world view promulgated in the Great Chain of Being. It is useful at this point to go back and review the historical roots of human spiritual experience.

Pre-cultures who lived at the dawn of human civilisation generated oral traditions that provided the first descriptions of the distinction between physical life and spiritual life. Shamanic space, which we have referenced, was a product of those very early clans and tribes, who pondered on life and death and sought to establish order in the minds of the living by affirming that there is birth, there is death, and the individual goes away. However, once a person weighs experience acquired personally, in solitude, the idea of going away is extended into a perception that there is a different place to go to. And that a return may be made.

This indicates that the messages presented via this twenty-first century transmission have been stated since the dawn of time. The primary variable has always been the particular imagination into which concepts were gifted. That has led to enormous variability in the detail of descriptions of life and death. Nevertheless, neither the parameters of life and death nor the cycle of incarnation have changed throughout time. Only the descriptions have changed. It is not that this or that historical tradition is right or wrong. That is not the point. It is rather that throughout human history recontextualised deliveries have been made available to those who were interested.

And so we reach the present time, which offers yet another opportunity to extend humanity's understanding of its existence. Of course, what happens after the delivery is made is entirely up to embodied humanity.

Our intention is to extend human understanding beyond the purely personal, and to anchor it within a useful model of existence that includes all life on this planet, not just the human.

The model of agapéic space presented here is an attempt to provide a rational model of all life on this planet. It supports the notion of equality that is not limited to the human but includes every species, given all species on this planet are bound in patterns of relationship. Of course, en masse species relate very closely—for example, minute plankton, which are fed on by larger magnitudes of creatures, are foundational to the survival of the planet as a whole. This is true of all species living at the bottom of each food chain.

Accordingly, if proper account is not taken of how life is interconnected, then human beings, who pride themselves as existing at the top of all food chains, live disconnected from their actual dependence on all other levels in the web of life. It is not at all that humanity is supreme, as some literature has proclaimed. It is rather that human beings are very fortunate that they have been positioned at the rop of the food chain for so long. That has not always been the case. And it is entirely possible that position will change again, for one reason or another.

This said, we now stand aside and allow that same identity who spoke in Chapter Three to comment on these vital issues, extending this discussion to consider humanity's responsibility within the web of life.

Humanity's responsibility

We begin by affirming that equality among all species is fundamental. This appreciation collapses the historically constructed hierarchical pyramid human beings use to view themselves as privileged and so having rights over all other species.

The only established right that applies in the context of the web of life is that any species seeking sustenance from a natural food source has the right to consume the target species.

Most species living in the web of life occupy a limited position in the food chain, and so have little choice regarding the species available for their consumption. In contrast, the omnivore naturally has broad reach, while humanity is renowned for its capacity to consume almost anything. Nonetheless, the breadth of humanity's reach does not privilege it in relation to food sources.

Instead, being top predator imposes an extraordinary burden of responsibility. Humanity is required to manage, in a balanced way, their consumption in order to protect every target species from which they derive nutrition. Formalised management strategies applied across the biosphere are now reasonably well established internationally. One could say these strategies have been adopted in the nick of time—except that humanity is already responsible for many species' decline and demise.

Historically, over-predation by a local human population has led to prey species becoming extinct. In most cases that local human population shifted onto other species, and so found sufficient sustenance to survive. However, in many places across the biosphere today an irrecoverable balance is developing between competitively predating humanity and key component species within the food chains on which human beings feed. Due to the looming collision of competitiveness and scarcity, excessive human consumption, and the consequent disruption of both environmental niches and foodchains, has become a threat to the planet as a whole.

Accordingly, we propose that a decline of the overall human population is now required. Humanity needs to reduce its own population to such a level that it no longer constitutes a threat to the populations of this planet's other occupants.

This is the responsibility human beings need to embrace. To successfully enact it, extended international oversight will be required, along with its sustained enforcemen. Implementing these will depend, in turn, on humanity utilising its own resources of wise counsel, whether that be achieved by appointing community representatives at the level of nation or state, or delegating upwards to the level of a world council. Whichever model is chosen, it needs to exist into the indefinite future and be sustained until humanity reduces its own population to such a level that it no longer constitutes a threat to all this planet's non-human populations.

In tandem with this requirement is the task of managing all species, ensuring populations do not fall to unsustainable levels.

In this context, we emphatically say to the world community that humanity must accept that because it is the most dominant it has the most responsibility in relation all other species.

The impact of human over-population

Humanity is out-breeding most other species. This must stop. An internationally adopted ethos in favour of population control is required.

Humanity lacks a control species to naturally limit, by predation, the burgeoning human population. Human beings well understand the principles behind population dynamics. Currently, there is an opportunity to apply that knowledge to their own species' limitation, by officially proclaiming that the fecundity of the human species exceeds the capacity of the planet to support it.

When the human population was one billion there was sufficient opportunity for all other species to survive and flourish within their environmental niches and positions in the food chain. This equally applied at all lower human population levels. At almost ten times that population there is now severe impact and resulting stress in numerous places across the globe.

One significant impact of human over-population is climate change. Altered temperature zones are pressuring many populations, human and non-human, forcing some species to migrate or die out. The magnitude of change humanity has wrought on this planet by unthinkingly maximising its dominance, invading other species' natural territories, and using technologies that multiply impacts due to energy consumption, is disrupting the web of life to a dangerous degree. The planet will take centuries to recover from what humanity has done. And is still doing.

If humanity possessed genuine concern for the rights of all other species, the only valid act would be to limit its own population to leave space for the natural balance of each and every species.

Who in good conscience would wish the opportunities for every other life form, both macroscopic and microscopic, to be so reduced that they have no space to survive, that their environment niches are ravaged or no longer exist, and that they have no chance to see out their life to its proper conclusion, either on the individual level or as a species?

Accordingly, we recommend that the same tactics used to constrain species that have become pests or harmful to human beings, such as releasing sterile members into a population to take up the activity of breeding with a known outcome, is exactly what is required in the human population. This will be controversial. We do not care. It is the only effective constraint.

Are you going to remain so blind, so self-focussed, so willing to bulldoze all before you, so stupid, greedy and disrespectful, as to render the entire planet a desert! The solution is obvious. Take the medicine. Be disciplined. Educate to achieve this. If you do not wish to educate, then sterilise by force, if necessary. For is that not what you already do to other species, exterminating all in your path? It will be an arduous and multi-generational task. But it is absolutely required. Techniques are available. They must be used.

A stark choice stands before you. You can have self-restraint or starvation. Which do you choose?

Suggestions for population control

An ethical human being will recognise the seriousness of the situation and, wishing to respect all other species, will find gentle and humane ways to restore the human population to a sustainable balance. We are in deadly earnest when we suggest that a population of one billion be the target. What is required is vigorous promotion of the zero-child family.

This will be unpopular, in the same way that Chinese imposed self-restraint has been unpopular. Yet it was tolerated at the same time, precisely because it was imposed from a political level, by top-down command. It will be similarly necessary for every other population group to accept this imposed discipline, strategically generated at the highest levels, possibly by the coming World Council.

Discussions concerning population that occurred in the 1970s, and falling birthrates in Western countries that followed, illustrate the potential impact if such ideas are promulgated internationally. Ethically-minded individuals will respond. The difficulty is that ethically-minded individuals could consequently be reduced to a prospective zero concentration in the resulting population mix.

Accordingly, it is insufficient to utilise only that technique to limit the human population. All the techniques known and understood in the fields of advertising and social control will need to be adequately funded in order to shift current opinion regarding human fecundity and make the prospect of having a child a sought-after privilege, not a right. If this smacks of eugenics, then let it. The situation is now far out of balance. Given the prospect that human population growth will continue, it is an issue that requires consideration, support and funding from the highest levels.

It will not happen easily. It will likely take two hundred years. But it is required, for the alternative is starvation, as we have said. Let that be sufficient motivation to drive the adoption of necessary policies, mandatorily imposed on the international community by a council of wise elders, given that most individuals have insufficient breadth of understanding and capacity for self-control.

This is a major ethical dilemma for your species. The Chinese decision will become lauded in the future as the first example of self-discipline in human population control. With that we end our environmental lecture.

Epilogue

HUMAN BEINGS DO NOT COMMONLY understand that from the perspective of the normal domain of humanity's existence, which is in agapéic space, going from there into physical existence could be described as a special excursion into a very strange place. It is a place in which one will be tested again and again as one attempts to gain some control of an alien monster, that being the physical body. In addition, as part of developing competence, one is also expected to negotiate alien physical territory and grapple with alien social terrain and alien moral compass.

Given the completely foreign nature of the physical body, at the outset individuals usually have little comprehension of its imperatives. Only after having occupied a body sufficiently that experiencing it becomes somewhat normal, and only after considerable exposure to the human social milieu, does one achieve sufficient understanding to temper and moderate the impulses that drive the physical organism, particularly impulses that come from one's own lower mind.

This perspective then raises the question: If this organism is so strange and foreign, what explicit benefits are obtained so as to justify this invasion of what could be construed as an alien mind? The justification is as follows.

The history of incursions into any foreign territory reveals that the challenges confronted and overcome lead to the acquisition of deep understanding, comprising morality, honesty, integrity, an ability to supervise others responsibly, and the potential development of skills in herd management, no matter which one. These benefits may be likened to those obtained as a result of travel.

When travelling you explore unknown territories, encounter different races, are confronted by foreign concepts, and need to cope with peculiar customs. All these radically expand your understanding. Your perspective expands from the narrowly parochial to the global. You develop acceptance of others who externally appear different from you. Your knowledge of the variety of cultures broadens. And you come to appreciate how diverse histories have led to cultural practices very different from your own. The sophisticated multicultured world citizen is generally believed to possess, and usually exhibits, the capacity to love all others without judgement due to a self-secure understanding of difference and an acceptance of codes of behaviour that previously appeared foreign, even derisible.

This is similar to what results after an identity, being a fragment of a node of Dao-consciousness, journeys repeatedly from, and permanently returns to, what is commonly called the spiritual domain. The identity will have encountered radically different behaviours, dealt with unanticipated consequences, and had its understanding stretched into unimagined dimensions, to a degree previously beyond its comprehension. That world traveller will have lived multiple lives in many locations around the planet, to speak only of this one. It will be imbued with a tolerance based on firsthand experience, having encountered a foreign culture not just for a few days, but having lived entire lifetimes within that culture. And then it will have experienced another culture. And another. And another. Each one different. Its understanding will have naturally achieved a condition that it is not surprised by any human action. It appreciates the embodied human throughout all possible experiences, the basis for its impassioned action, the range of its discursive thought, the intoxication of its emotional bliss, and the multiple limits that affect its capacity both to be violent and to love. All this so extends the understanding of the node of Dao-consciousness, and not just at the fragment level but at the level of the unified whole, that it achieves wisdom, given a wise person is one who has experienced much. On the basis that we are the product of that journeying, we would comment on the implications of the perspective just offered.

Just as it is nonsensical to bow down before anyone's house, so it is entirely inappropriate to define the zone occupied by non-embodied humanity as being so special it should be revered, so revered it should be considered the pinnacle of existence, and so positioned on that pinnacle that obeisance is made, so it is bowed down to and venerated in wonder and awe.

Yet this is in fact what is done by humanity, in all its fantasies of heaven. We now seek to level the playing field, to use that convenient metaphor. We wish to make clear that the movement identities make from the non-physical zone to the physical zone and back to the non-physical is lateral. There is no essential difference between the two zones. It is a level playing field.

This concludes our presentation of refreshed models and metaphors applied to human existence in order to enhance the understanding of you who seek a deep appreciation of what you are undergoing. You, as an embodied human, may be said to be the fragment of a node of Dao-consciousness. You have elected, for your own benefit, to join with, enliven, and eventually direct, the lower mind consciousnesses that exist within a series of human bodies. It is a process you will repeat until you have experienced all human life has to offer, and have learned to love equally, in any and all situations.

By loving equally we refer to treating others equally, applying the law equally, having equal expectations of all, behaving towards all equally, offering equal respect, and equally distributing knowledge of the parameters that support humanity's struggle to understand its proper place in relation to that zone from which all come. Practising equality also means seeing both the physical and non-physical zones as equal, and viewing the transfer from the physical to the non-physical as normal.

In stating this, we present an understanding of these things in a way that addresses historically accumulated misunderstanding of what may loosely be termed spiritual existence, and does so in a down-to-earth manner—which we mean metaphorically, of course.

This is our advice. Naturally, what is done is entirely up to each individual, which connotes equal respect on our part for each individual's free will. From this fresh beginning we wish the best for every intrepid traveller, that they may return from their journey safe and better informed.

Afterword: Peter Calvert and Keith Hill

Keith Hill

When I began my exploration of spirituality in the 1970s, conversing with non-embodied identities was not on my radar. In fact, for a long time I viewed channelled literature as unnecessary, even irrelevant, to spiritual communication.

However, meeting Peter Calvert in 2008 changed my outlook radically, because the material he was channelling was too informative, too stimulating, too *human*, not to be taken seriously. What he was doing also made me very excited. Peter's regular communication with non-embodied identities offered a great opportunity to seek answers to questions that I had long wondered about. Big questions, like how the universe came into existence, how human evolution occurred, and what consciousness is. So I formulated one hundred questions and Peter went away on a series of short retreats to channel answers. The entire process is presented in *The Matapaua Conversations*.

The Kosmic Web has developed out of that book. *The Matapaua Conversations* covers a wide variety of topics, often leaping from one issue to another within a single answer. Yet throughout it is very clear that behind the answers is a coherent world view, which would benefit from a different sequence of exposition. *The Kosmic Web* is the result. It reformulates selections from the *Matapaua* material, bringing disparate pieces of information into clearer relationship. It also adds new material channelled by both Peter and me.

That I now channel has been a wholly unexpected development. It came about through a series of steps. Originally, I offered to help Peter edit the channelled text for his second book, *Guided Healing*. While doing so I tried tuning myself to the guides' intent, so I wouldn't depart from or confuse what was being communicated. I did the same editing *The Matapaua Conversations*. The result was a realignment of my world view and an inner opening up to the guides' intent. After three years of this practice, one day while I was writing thoughts started coming through that were not my own. I have been channelling in written form ever since. The result is a still developing series of books.

Peter Calvert

My initial urge to meditate in a serious and prolonged manner occurred in 1990, when I began the Buddhist practice of Vipassana. I concluded that in 2010, after my tenth retreat. In 1999 I wished to expand my practice by meditating regularly with a friend, which continued for another five years. We subsequently slowly expanded the group size.

After meditating with friends for eight years, the idea came to establish a school to propagate the accumulated ideas derived from our channelling practice. Wisdomschoolnz was inaugurated in 2007 with a formal intention to continue to accrue the emerging material and to act as independent means by which to offer it to the community of those who are interested. That community was always expected to be small, but as the years passed, the scope of the material progressively expanded, both in relevance and application.

The tone of the material has always both intrigued and satisfied me. A mature and disinterested perspective devoid of drama is evident, occasionally pithy and amusing, while demonstrating astute insights into the human condition. As time went on, my understanding of the perspective offered—which views the human condition from a position consistently outside of it, a location I do not have—led to my eventual acceptance that this was a genuine spiritual teaching.

Now that a successful collaboration with Keith Hill has resulted in the production of a number of books addressing all of humanity, an expanded opportunity exists to share the material with those interested in their origins and destiny. To support that opportunity, I established a website— www.wisdomschool.nz —to enable international viewing of the original transcripts edited for anonymity and by topic. The published books are a much more coherent presentation of this material and its scope and relevance in the twenty-first century.

Glossary

Agapé: From the Greek, meaning selfless love. The word was subsequently adopted by Christian theologians to describe non-physical, spiritual love. Agapé is the nature intrinsic to Dao-consciousness.

Agapé frequency: One of three fundamental axes that define agapéic space. It has a scale comprising a nominal range of 1 – 65,000 discrete realms.

Agapéic space: A model and metaphor generated to describe spiritual existence. Agapéic space is defined by three axes at right angles to one another, these being agapéic frequency, hierarchy and willingness to bequest agapé. Agapéic space comprises the totality of existence, including the physical, astral and clear light domains. *See Air-fog model (in the Index).*

Ascension: Traditionally conceived as occurring vertically. Remodelled here as occurring within the axes of agapéic space and involving horizontal movement as much as, or more than, vertical movement.

Ask (of spirit): Spiritual identities commonly only communicate with human beings as a result of being asked. This enables them to avoid revoking any individual human being's free will.

Aura: The formative structure in the implicate order which partially defines the human physical body.

Auric channel: Refers to the aura's capacity to channel information from the spiritual domain to an incarnate individual's ordinary lower mind.

Awareness: A state of elementary or undifferentiated consciousness.

Axial Age: The period from 800 to 200 BCE, when new thinking about spirituality and reality occurred, as manifested in Taoism, Confucianism, Buddhism, Vedanta, Jewish prophets and Wisdom literature, Greek philosophers, etc .

Bio-identity: Identity at the physical human level, constituting a body, a biologically associated mind, and a personality. The bio-identity's mind is also called the lower mind. Co-associates with a Dao-identity.

Coalesence, co-association, co-habit: The process by which an individual spiritual identity unites with, and animates, a human body.

Channel: A human being who serves as the medium for transmission, facilitating the movement of information between the spiritual and physical domains.

Clear light: Spiritual space perceived beyond the dim astral level. Experiences of this domain have been placed on the historical record by mystics.

Consciousness: Consciousness comes from, and is a node of, Dao. It functions on two levels, Dao-consciousness and bio-consciousness.

Dantian: Historically, the Chinese term (*tan tien*) refers to an energy centre in the body located in the belly, approximately two finger widths below the navel. The Japanese term *hara* (meaning belly) is equivalent and commonly used interchangeably. The *tan tien* has been described as being "like the root of the tree of life". These terms refer to a different and deeper level of identity and function than the Indian term chakra, which relates to the aura. Within this text dantian is used to designate the kernel of the spirit-sphere when it is coalesced with its chosen body. *See Identity (spiritual)*.

Dao (Tao): The ultimate source of all that exists. The ultimate unmanifest.

Dao-identity: Identity at the spiritual level. In the case of the human, this is constituted of a fragment of a node of Dao-consciousness. It contains the qualities of the Dao, including intellect, purpose and self-creativity. Co-associates with a bio-identity.

Emergence: On the cosmic level, emergence is a feature of the expansionary cycle of each universe. On the biological level, emergence is a process whereby life as a whole develops from lesser to greater complexity. Biologically, this also applies to species. On the spiritual level, emergence is a process whereby nodes manifest from the Dao.

Empirical: Based on, concerned with, or verified by using sense-based observation or experience, as opposed to theory or pure logic.

Energy (particle): An electrophysical phenomenon measured in electron volts.

Energy (physical): The capacity to do work.

Energy signature (spiritual): A complex array of information encoded within the aura that conveys all of an identity's characteristics. Perception of an energy signature enables the perceiver to assess another's character and trustworthiness, and hence to evaluate the risk of interacting.

Expansionary impulse: A fundamental undefined cosmic phenomenon, causally attributed as a motive for species emergence, development and conclusion.

Great Chain of Being: A religious model of existence which depicts the hierarchical relationship of all creatures to God and the supposed divine realm, which is usually identified with heaven.

Group soul: A node of Dao-consciousness. Among those nodes that fragment in order to explore and mature, when they complete their cycles of incarnation they reunite and form a unified group soul.

Hara: Japanese for belly. It is an energy centre located approximately two finger-widths below the navel. As such, it is a patterning on the electrospiritual level derived from the implicate order. It is also the level on which the spirit-sphere manifests, and so may be interpreted as the level of actual human spiritual existence.

Haric space: Loosely used as an alternative name for agapéic space.

Hierarchy (social): A system or organization in which people are ranked above or below one another according to status or authority.

Hierarchy (spiritual): The attribute which constitutes the sum and product of loving acts throughout a life as one builds on one's history in other lifetimes.

Higher mind: A function of the higher self that processes spiritual information and manifests intellect and purpose. It is also where lower mind personality characteristics generated during incarnation are uploaded to and where those characteristics accumulate.

Higher mind's purpose: This is to integrate all gathered information, distil out repetition and redundancy, and come to clear mastery of everything associated with, related to, and implied by the accumulated information.

Higher self: The spiritual identity. It communicates with the lower self (that being the physical body and its brain) via the auric channel.

Human (conventional meaning): An individual consisting of body, social identity, mind and purpose.

Human (spiritual meaning): A spiritual identity, a node fragment of Dao-consciousness, who is either in an embodied or disembodied state.

Humanity: The entire set of humans currently embodied.

Humanity (steady-state model): This transmission proposes that an endless and continuous stream of spiritual identity is cast from the Dao, some of which elects to mature and become refined by engaging in a process of repeated incarnation within the crucible offered by human existence on this planet. This process is beyond the comprehension of the human mind, and so cannot be defined by, or limited to, the human concepts of past or future. This renders meaningless all issues concerning end-times and related eschatological phenomena.

Identity (spiritual): Modelled as an essentially invisible sphere with a small kernel at its centre. It exists as a structure in agapéic space where it is able to move at will, but to a degree dependent on its spiritual development.

Implicate order: A subtle underlying component of reality proposed by physicist David Bohm. Here identified as part of the electrospiritual. It manifests in the auric level structure necessarily associated with any living organism.

Individual (social): A single human being separate and distinct from others.

Individual (spiritual): A node of Dao-consciousness.

Intellect: That aspect of mind which processes information.

Intention: That process of mind which elects, evaluates and subsequently intends a specific outcome.

Karma: The outcome of human reactions which result in others being radically disempowered, confined or killed.

Kosmos: This word comes from the Greek and means ornament. It is used here in the sense that the kosmos is the ornament of the Dao, which manifested out of itself everything that exists. So the kosmos includes all domains, physical and spiritual, known, unknown and unknowable.

Light (physical): Electromagnetic radiation to which the human eye is sensitive, nominally of wavelength 380-720nm.

Light (spiritual): A subtle perception of brightness derived via the auric perceptual channel. Generally correlated with agapéic frequency, that is, dim at lower frequency and brighter at higher frequency levels.

Love: A human biological level emotion, in contrast to agapé, which is experienced at the spiritual level.

Lower mind: A function of the biological brain, it processes information and manifests habitual emotional characteristics that result from imprinting and enculturation.

Mind: See *Higher mind* and *Lower mind*.

Model: A hypothetical description of a complex entity or process. A representation of something, sometimes on a smaller scale.

Node: A zone of concentration of some attribute.

Node of Dao-consciousness: A condition of concentrated consciousness within the unmanifest absolute that contains intellect and purpose.

observer: The intent of the Dao that may be thought of as a large-scale emergent node of Dao-consciousness containing intellect and purpose. The unobserved initiator of the multiverse.

Point of attention: A product of focussed awareness, free to move without limit of space or time.

Purpose: See *Higher mind's purpose*.

Realm: This transmission proposes, in its model of a agapéic space, that there are 65,000 discrete levels of agapéic frequency. Each level is a realm that provides a possible location for a node of Dao-consciousness comprising spiritual identity, and typically, for a population of such nodes.

Religion: An informal or formalised organisation providing social engagement in a common interest. That interest is normally directed towards non-physical existence in order to contextualise human life.

Sensitivity: The ability to respond to stimuli or to register small differences in stimuli.

Sentience: The readiness to perceive sensations. Also indicates an elementary or undifferentiated consciousness.

Soul (group): In undivided form, the node of Dao-consciousness. In divided form, a cluster of fragments, more commonly known as individual souls or spirits, each of which incarnates individually in order to gather information. At the end of the cycle of incarnations the fragments re-unite to form a mature and refined node of Dao-consciousness. This transmission comes from such a re-united group soul.

Soul (individual): A traditional religious term. Here it is redefined as the fragment of a node of Dao-consciousness. This node fragment manifests as and is comprised of a spirit-sphere that co-associates with a human animal body, coalescing with it and thereby animating it.

Spirit: See *Soul* and *Spirit-sphere*.

Spirit sphere: An individual fragment of a node of Dao-consciousness in globular form. In human beings it manifests at the hara-level structure and defines individual human identity. It may only be observed via non-physical visual perception using the auric channel.

Spiritual identity: A node or node fragment of Dao-consciousness.

Spiritual realm: The Dao, including all of manifest and unmanifest existence.

Spirituality: The idea or act of attending to a generally invisible domain of reality. Historically, it has been defined in different ways by different religions and cultures.

Survival: When a physical body dies its associated lower mind dies with it. What survives is the accumulated information gathered during incarnation, and the spiritual self and its higher mind to which that information is uploaded.

Teaching (spiritual): A knowledge-set deliberately transferred from identity in spirit to lower mind in body.

The Real: An historical spiritual conception that proposes that the spiritual domain is more significant and real than the physical domain. That contention is not supported in this teaching.

Time (disrupted or missing): Accessing the higher mind's level of awareness is normally associated with apparent disruptions or discontinuities in the ordinary awareness of passing time. This is a consequence of the functioning of bio-consciousness, not of Dao-consciousness.

Transmission (spiritual): A spiritually initiated transfer from a higher mind to a lower mind, usually comprising knowledge of spiritual existence, of physical existence from a spiritual perspective, and the relations between them.

Index

To the reader

Small presses rely on supportive readers to tell others about the books they have enjoyed. To support this book and its authors, we ask you to consider placing a review on the site where you bought it.

Are you interested in learning more about the guides' perspective on humanity's spiritual nature and development? A number of avenues are available. Peter Calvert has archived all the material he has channelled since 1998. This is a huge resource available to be freely viewed online at wisdomschool.nz. In addition, on the website www.experimentalspirituality.net the guides offer observations on a range of topics.

At the time of this second edition, in 2022, Peter Calvert and Keith Hill have produced twelve channelled books—with several more in development. Initiated by the guides, these books offer a generally consistent approach to individual spiritual development. The channelled books Peter and Keith have produced to date, together and separately, divide into several categories:

METAPHYSICS FOR THE TWENTY-FIRST CENTURY
— Channelled by Peter Calvert and Keith Hill

Metaphysical descriptions map the unknown terrain that surrounds us. Historically, religions have supplied humanity's metaphysical descriptions of reality. But today they have become outmoded. We need new metaphysical descriptions that chime with current knowledge. Such descriptions are offered in the following books.

The Matapaua Conversations

This book is Peter and Keith's first together. Keith asked the guides one hundred "big" questions regarding the big bang, evolution, the nature of consciousness, and other related metaphysical topics. Peter then took time out at Matapaua Beach to receive the answers, keeping a diary to record his experiences and thoughts throughout. This book complements the material presented in *The Kosmic Web*.

Learning Who You Are

Drawing on a variety of texts, including Peter's extensive archive of channelled material, this book offers an easy-to-read overview of the guides' outlook. This is an ideal introduction to the guides' outlook.

CHANNELLED BY PETER CALVERT

Guided Healing

This book contains two urgent messages. To spiritual seekers, *Guided Healing* presents a novel view of the spiritual purpose and benefits of being born into a physical body. Issues covered include the relationship between the spiritual and physical realms, the reason for incarnation, the use of meditation as a means for exploring the spiritual realm, and the significance of soul work.

To healers, *Guided Healing* offers instruction on how to become a conduit for healing energy that emanates from the spiritual realm. Topics covered include how to contact guides in the spiritual realm, the nature of spiritual perception, and factors which enhance or hinder energy flow during the act of healing.

Agapè and the Hierarchy of Love

After years of meditating, Peter Calvert found himself communicating with non-embodied beings. These beings gave him metaphysical and personal training in what they called "spiritual empiricism". This book is a compendium of these messages. It includes numerous models that illustrate the relationship of the physical and spiritual domains.

CHANNELLED BY KEITH HILL

THE CHANNELLED Q+A SERIES

This series presents easy-to-read, non-technical books that explore spirituality at an introductory level. For each, the guides set a general topic for discussion, then had Keith Hill, this series' channeller, invite people he knew to submit questions, which the guides then answered. Each book contains twenty-one questions and answers, with a number of questions picking up and expanding on prior answers. The result is a spontaneous to-and-fro in which, as the guides comment, many surprising and unanticipated topics are explored. On the other hand, the guides also surprise with their answers. As of this publication there are three books in this series. While they are listed in the sequence they were created, they are designed to be read as stand-alone texts, and in any order.

What Is Really Going On?
The first in this series focuses on reincarnation and its implications, along with a range of key spiritual topics.

Where Do I Go When I Meditate?
The guides discuss not just the possibilities that meditation creates, but also chakras, prayer, what we gain from entering the spiritual realm, and the nature of extra-terrestrials—among which human beings must be numbered.

How Did I End Up Here?
This book addresses life plans, diet, allergies and autism, personal development and introduce the process of self-enquiry.

THE CHANNELLED SPIRITUALITY SERIES

This series is designed for those who wish to engage in personal development. It provides a means for seekers to understand their individual psychospiritual make-up, what factors drive their current existence, and what the key factors of their life plan involve. The books in this series together offer a practical way to carry out self-enquiry, whether in groups, solo or in professional contexts.

The series' channeller, Keith Hill, has a background in the Fourth Way Work teaching of G.I. Gurdjieff. Because these books have been offered through Keith's mind, they build on what he understands of the Gurdjieff Work. In addition, Keith has been directed to the Michael Teachings. These were originally channelled by a group in San Francisco, who had also been trained in the Gurdjieff Work. The Michael Teachings add many details to the Gurdjieff Work, the most fundamental being that it puts psychospiritual development into a reincarnational context. The Michael Teachings also add considerably to Gurdjieff's ideas on human psychological make-up.

Selections from all this material have been utilised by the guides in the

Channelled Spirituality Series to create a straightforward, psychologically-based approach to personal psychological and spiritual development.

Experimental Spirituality

Introduces the rationale for adopting a non-religious, empirical and experimental approach to spirituality. Key concepts include the journey from belief to knowledge, how human identity is structured, the nature of the five-layered self, and how developing understanding relies on asking questions in the right way.

Practical Spirituality

Considers the psychospiritual factors we draw on when planning each incarnation. Topics include examining reincarnation and its impact on evolving identity, the rationale behind life plans, the nature and purpose of karma, confronting negativities and nurturing positive qualities within, and the role of life lessons in helping us mature.

Psychological Spirituality

Uses the paired concepts of true and false personality to explore how human identity is formed and plays out during the course of a life. This book also considers how lives are linked into sequences that form a unique trajectory through the human world, and what it takes to initiate self-transformation.

All books may be purchased online and from your favourite bookstore. To learn more and to read chapter excerpts go to www.attarbooks.com.

CPSIA information can be obtained
at www.ICGtesting.com
Printed in the USA
BVHW030902230123
656893BV00013B/476/J

9 781991 157027